THE
PLANETARY ATOM
A FICTIONAL ACCOUNT OF GEORGE ADOLPHUS SCHOTT
THE FORGOTTEN PHYSICIST

THE
PLANETARY ATOM
A FICTIONAL ACCOUNT OF GEORGE ADOLPHUS SCHOTT
THE FORGOTTEN PHYSICIST

by

Jean-Patrick Connerade, *alias* Chaunes

Former Lockyer Professor of Physics
of the University of London
at Imperial College

Foreword by
Roald Hoffmann, Nobel Laureate in Chemistry

Translated from the French original
"L'Atome planétaire"

 **World Scientific**

NEW JERSEY · LONDON · SINGAPORE · BEIJING · SHANGHAI · HONG KONG · TAIPEI · CHENNAI · TOKYO

Published by

World Scientific Publishing Europe Ltd.

57 Shelton Street, Covent Garden, London WC2H 9HE

Head office: 5 Toh Tuck Link, Singapore 596224

USA office: 27 Warren Street, Suite 401-402, Hackensack, NJ 07601

Library of Congress Cataloging-in-Publication Data
Names: Connerade, J. P., author.
Title: The planetary atom : a fictional account of George Adolphus Schott the forgotten physicist /
    Jean-Patrick Connerade alias Chaunes, Former Lockyer Professor of Physics of the University
    of London at Imperial College ; foreword by Roald Hoffmann, Nobel Laureate in Chemistry ;
    translated from the French original "L'Atome planétaire."
Other titles: L'Atome planétaire. English
Description: New Jersey : World Scientific, [2022]
Identifiers: LCCN 2021021559 (print) | LCCN 2021021560 (ebook) |
    ISBN 9781800610026 (hardcover) | ISBN 9781800610149 (paperback) |
    ISBN 9781800610033 (ebook) | ISBN 9781800610057 (ebook other)
Subjects: LCSH: Schott, G. A., 1868-1937. | Physicists--Biography. | Quantum theory. | Atoms.
Classification: LCC QC15 .C67 2022 (print) | LCC QC15 (ebook) | DDC 530.092 [B]--dc23
LC record available at https://lccn.loc.gov/2021021559
LC ebook record available at https://lccn.loc.gov/2021021560

British Library Cataloguing-in-Publication Data
A catalogue record for this book is available from the British Library.

For any available supplementary material, please visit
https://www.worldscientific.com/worldscibooks/10.1142/Q0294#t=suppl

Desk Editors: Balasubramanian Shanmugam/Michael Beale/Shi Ying Koe

Typeset by Stallion Press
Email: enquiries@stallionpress.com

Printed in Singapore

*T*he *Planetary Atom* is a wonderful, gentle novel of a place and time and people, that seem light years away from the posh settings and grant-angst-ridden denizens of contemporary laboratories. A time when scientists listened to each other with respect, even when they patently disagreed. Letters were answered in days, there were weeks between conversations. And the august institutions of science didn't look at citation indices.

We romanticize that time, always set in the past (while imagining a better future). Yet it is always hard to find. If there ever was such a time — and this book constructs it convincingly — it is clear that it was created by special people. Thus, Ernest Rutherford, a forthright New Zealander — 'a force of nature' — as Paul Langevin called him, could give George Adolphus Schott, a brilliant mathematician junior to him, the time of day. To listen to Schott's explicit, even if most politely worded, disagreement with Rutherford and Bohr on the planetary atom helps create that emotional space of a time when physics gathered pell-mell around another viewpoint. Clearly, Schott found a way to do good science at Aberystwyth and the Royal Society had the good sense to elect him to membership with sixteen publications to his credit.

The novel is George Schott's story, a story of an unassuming scientist building a world out of mastery of electromagnetic theory. We see how he gained the peace of mind to allow him to continue thinking; and how, in some way, it did not matter that it was ten years after Schott's death that a kind of radiation he had predicted would be recognized. It takes a special author to deliver a story which understands that there is room for everyone.

Roald Hoffmann
Nobel Laureate in Chemistry

L ittle is known about the lives of some very important scientists. The most significant scientists are, of course, those who discover previously unknown laws of Nature or who formulate new ideas which subsequently structure our thinking. When such individuals disappear from the scene without leaving any key to their discoveries, one cannot help feeling frustrated by not fully comprehending the path they followed to uncover profound truths.

Such ignorance is understandable, for example, in the case of Democritus, who invented atoms such a long time ago. It is already miraculous that we actually know about him and his idea today. A proper biography of the man would be asking too much for such a remote period in history.

However, it can also happen that information has been lost about a more recent researcher, one who contributed in a fundamental way to science and yet attracts less attention than he should, either because his contribution was not accepted, in his own times, or because the discovery came at the wrong moment or in the wrong place. Thus, Ignác Fülöp Semmelweis understood the importance of doctors washing their hands well before Louis Pasteur, but did so at the wrong time in history. Similarly the librarian of Alexandria, Eratosthenes, accurately measured the circumference of the globe in 200 BC, without dreaming that, for hundreds of years after him, many Europeans would still believe the Earth to be flat and, even more surprisingly, that the children of those barbarians from the north would write the history books much later, to celebrate their own circumnavigation of the planet.

It is more troubling when almost nothing is known about an important innovator of modern times, not only because few have actually heard about his work, but also because he was so shy and discrete that he went out of his way to shirk any form of publicity.

Such is the case of George Adolphus Schott. In fact, we know so little about him that the best option is to reinvent his biography, using his scientific contribution to guide us as to what might have been the context and the important events of his life. This is no longer the work of a scientist or of a historian. It becomes a subject for a novelist, because it rests on a complex combination of imagination, psychology and understanding of how scientific advances really occur.

Why even pursue such a path? The answer is that Schott was a great scientist who never achieved proper recognition and that he remains in desperate need of rehabilitation. Somehow, he has been pretty well erased from the history of modern physics, despite having played a crucial role. So, we do owe it to his memory to try and imagine what his life was like in order to restore him to his rightful place.

Since so little is known, the best possible course was indeed to make his story the subject of a novel.

Many will say: science is science and works of fiction are works of fiction, implying that the two worlds should never meet.

However, this is not quite right.

Writers and poets rarely take an interest in physics. When they do, they are liable to come up with the kind of glib rhetoric they know in advance will satisfy their public, in the style of Edgar Allan Poe's much loved (and too often quoted) 'Sonnet to Science'. Few people remember that Poe is also the author of *Eureka*, but maybe the public also likes poets to be consistent in their rejection of natural philosophy. So, Poe's anti-science sonnet is almost always regarded as his considered view.

Is this sensible, however? Should one prefer Byron's hatred of modern science to Shelley's fondness for the explosions and horrible smells of chemistry? Are the poets allowed to have an independent opinion, or has the public decided on their behalf to exclude the sciences from matters deemed poetic?

The concept of a work of imagination which puts science at its heart and treats the selfless pursuit of scientific investigation as a human passion equal to others in life is unusual. It is aimed, not only at readers of novels, who do not usually tread such ground, but also at practitioners of the 'scientific literature' which rarely contains anything remotely literary or poetic within its pages.

This book is a celebration of science coming from an allegedly opposite camp. It attempts to view the pursuit of science from the inside, as a form of artistic expression. Perhaps that is why the author chose the discipline with the worst possible reputation amongst artists in general, namely physics, queen of the hateful 'exact sciences' and, within this subject, the birth of quantum mechanics. No area of science might be expected to strike greater fear into the hearts of poets and their readers.

Can science inspire, not only those who practise it in their own lives, but also the artists and poets whose taste in intellectual adventure does not usually lead them into the land of mathematical physics? Usually, acceptance of science and its certainties is argued as some kind of modern necessity, a reality one must come to terms with in order to qualify as a true citizen of our times. Falling in love with uncertainty and deep forms of conjecture does not sit comfortably with this conventional view.

The present novel does not claim to make science seem useful. It has no practical agenda of any kind. It does not seek to persuade the reader that ideas the characters are so passionate about might have applications. Their quest for truth is a mixture of idealism and respect for the deeper significance of Nature. It has nothing to do with the business of making things or of contributing to the well being of the modern world. The discoverers themselves, see the realities of industry and trade as secondary compared to the beauties of fundamental science.

Such love of pure science is in no way connected to motivations expected by society. If truth be told, our leaders and their advisors are more interested in innovation than in science itself. To the world at large, the scientist described here is almost as useless as a poet. Even if he does discover something 'worthwhile', the likelihood is that he will not know what to do with it or how to exploit it. In the modern world, a researcher must be followed around by somebody with sound practical sense, who understands where novelty becomes profitable and will turn it into a money-making activity, justifying investments made by the State. The governance of science was invented to protect us all from 'mad scientists' who pursue research for reasons clear only to themselves. Science is then useful, and commendable as a pursuit.

However, this form of justification is anathema to the poet. He is poles apart from the adage of George Porter: 'there are only two kinds of science — applied science and not-yet applied science'. To him, indeed, the merest hint of application signals that the time has come to move on and quit such a deeply distasteful activity.

Fortunately, what is recounted here dates back to a time when researchers were not faced with this constraint. The concept was, in those golden years, that science in its purest form is quite the equal of art and should be thought of in precisely the same terms. In those days, great scientists, like great artists, pursued a dream. The mystery of such dreams, how they arise and how they inspire, is ultimately the subject the author has chosen to address.

Jean-Patrick Connerade

'I think a strong claim can be made that the process of scientific discovery may be regarded as a form of art. This is best seen in the theoretical aspects of physical science. The mathematical theorist builds up on certain assumptions and according to well-understood logical rules, step by step, a stately edifice, while his imaginative power brings out clearly the hidden relations between its parts. A well-constructed theory is in some respects undoubtedly an artistic production. A fine example is the famous kinetic theory of Maxwell... The theory of relativity by Einstein, quite apart from any question of its validity, cannot but be regarded as a magnificent work of art.'

Ernest Rutherford

Jean-Patrick Connerade is currently Emeritus Professor of Physics and Distinguished Research Fellow at Imperial College London, President of the European Academy of Sciences, Arts and Letters, Honorary President of EuroScience, Honorary Professor of Physics at the East China University of Science and Technology in Shanghai, and Permanent Guest Scientist at the Wuhan Institute of Physics and Mathematics of the Chinese Academy of Sciences. He is a former editor of the *Journal of Physics B* (UK), author of the book *Highly Excited Atoms* published by the Cambridge University Press (translated and published in Chinese by the Chinese Academy of Sciences), Fellow of the Royal Society of Chemistry and Fellow of the Institute of Physics.

As an author, Jean-Patrick goes by **Chaunes**. **Chaunes** is a poet, author and playwright in French. He has won several prizes for his poetry including the Grand Prize for Poetry (Victor Hugo Prize) of the Society of French Poets; the José-Maria de Heredia Prize of the Académie Française; the Paul Verlaine Prize of the Maison de Poésie; the World Prize of Humanism; the Gotseva Misla Prize of Literature and many more. He is an Honorary Member of the Société des Poètes Français, Member of the Société des Gens de Lettres, and a Member of the Académie des Sciences, Inscriptions et Belles-Lettres de Toulouse.

Published by EuroScience, Strasbourg, France:

Chaunes and Wolfgang Heckl, *Ein Visionär auf dem Bayerischen Thron — Ludwig II the visionary King of Bavaria* in Munich in 2006, ISBN 978-2952720002.

Chaunes (Ed.), *Science Meets Poetry*, Proceedings of the International Meeting at ESOF in Barcelona in 2008, ISBN 978-2952720021.

Chaunes (Ed.), *Science Meets Poetry 2*, Proceedings of the International Meeting at ESOF in Turin in 2010, ISBN 978-2952720038.

Chaunes and McGovern (Eds.), *Science Meets Poetry 3*, Proceedings of the International Meeting at ESOF in Dublin in 2012, ISBN 978-1481951005.

Chaunes (Ed.), *Science Meets Poetry 4*, Proceedings of the International Meeting at ESOF in Copenhagen in 2014, ISBN 978-1505281538.

Chaunes and Illingworth (Eds.), *Science Meets Poetry 5*, Proceedings of the International Meeting at ESOF in Manchester in 2016, ISBN 978-1542436557.

Chaunes Dazet-Brun and Dorge (Eds.), *Science Meets Poetry 6*, Proceedings of the International Meeting at ESOF in Toulouse in 2018, ISBN 978-1796671278.

Translations into English published by aux poètes français:

Shakespeare in Fulham (a play), ISBN 978-1686197697.

The Many Lives of Ivan Kazanovich (a novel), ISBN 978-1977629968.

And other books in French under the pseudonym 'Chaunes'.

Scientific works under the name Jean-Patrick Connerade include:

Connerade, J-P., *Highly-Excited Atoms*, Cambridge University Press, 1998.

Connerade, J-P. (Ed.), *Correlations in Clusters and Related Systems*, World Scientific Press, 1996.

Connerade, J-P. (Ed.), *Latest Advances in Atomic Cluster Collisions*, Imperial College Press, 2008.

Chaunes

Grand Prix Victor Hugo de la Société des Poètes Français

Prix Jose-Maria de Heredia de l'Académie Française

Prix Paul Verlaine de la Maison de Poésie de Paris

World Prize of Humanism Ohrid Academy

Gotseva Misla Prize of Literature Skopje

Contents

Scientists Versus Poets?

There exists a quite mysterious connection between the destiny of great scientists and the lives of famous poets. Both experience a period of public doubt about their own worth which in some cases can be long and seems a necessary step before recognition. This period is important: it is usually the moment when their immediate entourage begins to appreciate the full significance of their achievements.

In this respect, science resembles high forms of poetry which are only practised, as a rule, by a small number of exceptionally gifted individuals. It is often said that only five people in the world could really understand the great scientific papers of Albert Einstein when they were originally published. If such a statement can be taken seriously, then one of the five was surely George Adolphus Schott. He had read all of Einstein's papers, including, very importantly, those written in the German language.

Recognition, however, in the case of Schott, followed a long, unfair and tortuous path. It is not enough to display remarkable talents. It is also necessary to attract attention. From the outset, that was the problem faced by George Adolphus Schott.

The similarities between the scientist and the poet are clear enough, but the differences are also fundamental. It is much more likely that a man of science will be recognized as such within his own lifetime. In fact, it is important for him to achieve that status as quickly as possible, because science itself moves on, and does so very quickly. The present times, in research, are very short-lived. As it moves ahead, the investigation of truth destroys its own past and is under constant threat of being swept away, along with old reputations, swiftly consigned to oblivion. So, a man of science has a short time to make a name for himself and his window of opportunity quickly disappears.

The poet, on the other hand, is unlikely to achieve recognition within his own lifetime and, if he does, should be wary of such early success. It is a potentially dangerous kind of fame which might not last for long.

How is it that the scientist can enjoy the advantage over the poet of experiencing celebrity? There seem to be two reasons. The first is that science (it is often said) is built on certainty. This may or may not be quite true but, without doubt, experiment plays a marvellous and irreplaceable role in the confirmation of new ideas. Often, one has the impression that Nature itself comes forth to answer the great questions scientists put to her. She emerges from her secret depths only at the most important and legendary moments in the history of science. It could be in the observations of an eclipse by an expedition to the other end of the world. It might be when a telescope borne by a rocket sends back signals through space about the properties of the universe. On such occasions, Nature confirms the most profound discoveries made by man and dismisses all the worrying ambiguities of speculation.

Poets must live without this magnificent advantage. A poet is alone with his public and Nature sends him no signal to confirm or refute his poetry. He also invokes Nature but the dialogue is one-sided and there is not much purpose in calling her to his rescue as she will not come.

The second disadvantage of the poet is sociological. Scientists form a powerful lobby. Collectively, they shape opinion and even the most powerful statesmen must heed their voice. Who knows what they might think up next? Are they about to change the world again and transform all the habits of our society? It is better to be wary of them as a group and a politician is usually reassured by their approval. The physicists and chemists, in particular, were amongst the first, together with medical doctors, to create powerful professional organizations and throw their weight around in the political arena. They made use of the Nobel awards as an instrument to achieve notoriety, and the recipients of this famous prize could lay an almost unquestionable claim to fame and glory, to the extent that nations jockeyed for their turn in lobbying for the prestigious award.

Against this, the men of letters, and especially the poets, had far less success. As regards poets, prizes bring no security. Someone completely unknown might well step out of the shadows after the demise of prestigious prizewinners and rob them of the leading role they had fought so hard all their lives to conquer. Poets must live with this kind of uncertainty. Nobody can tell what their future will be made of.

Not only can scientists become famous in their own times. The same is true of the walls they work in. Their universities, their laboratories, the places where great discoveries were made become celebrated institutions which command respect as though there were something magical about the very location where it all happened.

In this respect, the most famous laboratory for physicists is undoubtedly the Cavendish Laboratory in Cambridge. In its golden years, it established itself as the most important institution in the history of the exact sciences. No other laboratory in the world can claim to have produced quite so many Nobel Laureates per square metre and to have done so in so short a time.

The leader of this mythical laboratory, Ernest Rutherford, was a kind of intellectual Midas. All those who worked around him, students and collaborators alike, every member of his team seemed destined sooner or later to collect a Nobel prize. This was a unique and astonishing situation, too improbable ever to be repeated anywhere else or in the future.

All of them, it must be admitted, except one.

There had to be at least one exception to the rule. After all, science itself is a human construction and everybody knows, since biblical times, that human constructions are of necessity imperfect. So, there had to be, behind the ideal facade of the Greek temple of scientific research, hidden between the columns in some dark corner, a representative of the human condition, some more frail character who never achieved the prominent place expected of him and might, in the midst of so much fame and glory, appear as a kind of scientific black sheep, one with a destiny closer to that of the poet. The tale of the odd one out amongst those who surrounded Rutherford, the forgotten star of the University of Cambridge, is truly fascinating. It teaches us some very real lessons about the vanity of reputations (even: of scientific reputations) and about worldly recognition in general. That is good reason to visit his story and remind ourselves of his contribution to science which, oddly, is only known today to a rather small number of specialists.

Why is a distinguished scientist all but forgotten today? Scientists, like poets, are sometimes born in the wrong century for their talents, or in some unsuitable location, or in the wrong social class for study or, again, in a family whose beliefs do not favour an open mind. History provides plenty of examples of such impediments. With regard to poets, one thinks of Constantine Cavafy, born on the very periphery of Greek culture, in a period desperately far from the times of Homer or the conquests of Alexander. Nonetheless, he overcame all these obstacles to become a great voice of Hellenism in modern times.

In the sciences, searching for comparable examples, the first case that would spring to mind is indeed that of George Adolphus Schott. He came forward to propose a profound advance in human knowledge, but nobody wanted to hear about it there and then. Despite his acknowledged brilliance, the idea seemed to run counter to the spirit of his time. Some thought it even contrary to progress. Worse still, it emerged later that it had been thought of too early and that the systems to which it could be applied did not yet exist. They had not been discovered or assembled in the laboratory. So, another scientist came forward. He stole

the idea and, either willingly or by accident, pushed Schott's work into historical oblivion.

Yet, his career had begun quite brilliantly. Nothing suggested such a perverse twist of fate. In fact, his story stands as a kind of inverse march away from glory which gives him the aura of a scientific Don Quixote of modern times. It was his destiny to tilt against windmills and always to pick the losing side which he would invariably serve with amazing ability.

The fact is that he was often right in what he argued, but always did so at the wrong moment in time, either too early or too late, both of which are usually fatal.

There is a strange kind of greatness in missing so many important opportunities. If one can assess his life correctly, it is probably true that Schott should have found a leadership role worthy of his talents. However, since there was none on offer, the only place left for him was a romantic and fascinating role as the misunderstood and forgotten genius of his generation. His contemporaries did not detect his importance and did not know what to make of his work. So, he missed the opportunity of being recognized. As a result, he remains in obscurity even today, unless one makes the effort to explore the past and discover him. The effort, however, is well worth it: often, the best mysteries are those which lurk in shadows where nobody is looking.

2

Ernest Rutherford Meets George Schott

When George Adolphus Schott appeared for the first time before him, even Ernest Rutherford was impressed. Schott, it must be said, was one of the most brilliant scholars of Cambridge University, a former student of Trinity College — a real Mecca for mathematicians. He was a Senior Wrangler and there seemed to be no distinction he had failed to win from the university and college where Rutherford himself was proud to have completed his studies. Schott was renowned as the most excellent of mathematical physicists and (last but not least) captained the prestigious Chess team of Cambridge University.

Even Rutherford, despite his alleged distrust of theoretical physicists, could not fail to recognize the importance of meeting George Schott. He had before him one of the great scientific minds in England, whose reputation as something of a mathematical genius was already well established among his peers and left little to be desired. What also struck him immediately was the contrast between Schott's manner and style and his own nature. The two men were very different from each other and the contrast between them would work its way through even down to their approach in unravelling properties of Nature. Rutherford sensed immediately that Schott's thinking was somehow complementary to his own and that the two of them, working together, might form an irresistible tandem.

Rutherford knew him already by reputation although he had never met him before: Schott was his elder by three years and had drifted through Cambridge like a comet in the sky of mathematics in the very same department as the New Zealander was to complete his studies five years later. Both of them were precocious scholars. However, Schott's reputation was essentially confined to the small circle of former students at Trinity College, Cambridge. Rutherford, on the other hand, had already been appointed professor at the age of 26 in the McGill University of Montréal, Canada, and then full professor in the University of Manchester. He was a rising star with a worldwide international reputation for his discoveries. He was also a prominent speaker, with a booming voice and a strong presence, while Schott was a quiet and rather diffident man, very elegant in his ways but always exceptionally discrete when expressing himself.

Despite that, Schott's legendary ability had left a strong mark in Cambridge, and Rutherford, of course, was well aware of his remarkable accomplishments.

There was, however, one respect in which Schott's career seemed very enigmatic to Rutherford, and their meeting did little to dispel the mystery. After such a spectacular start as a mathematical physicist, why on earth had he withdrawn to an obscure new provincial school, hardly even a University, such a small and unimportant place? It seemed an odd choice, a kind of eccentricity on the part of a man whose achievements would have been the envy of any budding scientist. Nobody had even heard of the University of Aberystwyth. Even the name seemed rather absurd, and Rutherford, as a New Zealander anyway, read it several times over wondering about its pronunciation.

In truth, it was a very second-rate institution. It had only just been created in 1872 and had basically no history in any subject worth speaking about. What was Schott doing in such a place? It was also a mystery amongst the former scholars of Trinity. Why even consider a university in Wales? The whole idea seemed totally absurd.

One could perhaps interpret it as a strange form of humility. Coming from a man who, until then, had always achieved spectacular success without ever pushing himself forward or blowing his own trumpet, it was perhaps a signal that his moral values were not quite the same as those of those who always try to push others aside and climb to the top of the ladder. Maybe it was a way of expressing that he did not expect to reap any reward in terms of position until he had accomplished some truly important advance.

Perhaps it was the attitude of a real gentleman, a person who felt it vulgar to overclaim and to expect rewards merely for passing examinations with very high marks and understanding difficult branches of learning.

As a farmer's son, and a young professor recently returned from abroad and accustomed to rather brash ways of making his mark, Rutherford was somewhat taken aback by this aspect of Schott's personality. In fact, he rather misinterpreted it: he attributed Schott's shyness to some kind of complicated aristocratic background. Did he belong to the strange group of students from high-class families Rutherford had occasionally met in Cambridge, without ever understanding what they were doing there? Such people regarded themselves as important simply by birth. Perhaps Schott's family also belonged to the Establishment. Maybe he had no real need to study and science, for him, was just a kind of hobby, not to be taken too seriously.

Rutherford had seen such people and knew they existed, although he had no idea what really motivated them.

In fact, Rutherford's supposition could not have been further from the truth. George Adolphus Schott came from a fairly prosperous middle-class family, but had no connection whatsoever with the elite. True, his parents were wealthy enough to employ a few servants, but they were not rich in the real sense of the word. Schott's character made him elegant and discrete, which spoke in his favour and might have created an impression of distinction, but there was nothing aristocratic in his background at all. His style of dress was difficult to place. It was not even completely English. Added to which, he spoke too well, followed social rules too precisely, was, in a sense, too well behaved. During his time as an undergraduate, he was even railed for mannerisms by his peers as a student. The word went round: 'He's just too English to be true' — a rather cynical assessment of his character, but so accurate that it stuck.

George Adolphus Schott was in reality the son of German immigrants to the United Kingdom. He usually presented himself simply as 'George' and dropped the 'Adolphus' as he was conscious the second name sounded foreign and even a bit ridiculous in England. Later on in life, he even tried to erase his second name and looked for another when it became a further embarrassment to him for political reasons. What made him so shy about his own identity was psychologically rather complicated. He was always careful not to reveal his German connections except to intimate friends. Disclosing his second name became a sign that he really trusted them. Schott was rather secretive, quite the opposite of Rutherford, although both of them, in a sense, had immigrant backgrounds. Rutherford, on the other hand, was very proud to be a New Zealander and always said so.

On their first meeting, although Rutherford had the undeniable advantage of being famous, he did feel slightly inferior socially, for reasons he could not fathom precisely. Next to him, Schott had the style of a man of the world, an upper-class gentleman with the kind of social veneer needed to shine in Victorian England. He had elegance, the mysterious quality of the middle classes, perhaps inherited from some kind of family tradition.

Rutherford was destined to become the leader of the Cavendish Laboratory in Cambridge and President of the Royal Society, but he always treated Schott with great respect after this first meeting. It was not so much because of his mathematical talents or his reputation as a chess-player. It was a question of style. As the son of a New Zealand farmer, Rutherford lacked certain social skills. He knew how important they were in English society and Schott seemed to display all the talents he could not master.

Both were former students from Trinity and, although their paths had not crossed earlier, they had common memories of tutors and teachers with odd and amusing habits. Both remembered the theologian who kept a bottle of port hidden behind his books and the lecturer in mathematics who carefully wrote out all his jokes in the lecture notes and repeated them unchanged, year after year. Such tales belong to the folklore of students and teachers alike and soon intruded in their conversation.

Gradually, they worked round to the real subject of their encounter, the deeper motivations which had driven George Schott to Manchester for his first meeting with Ernest Rutherford.

'It is because of your Planetary Atom,' said Schott. 'I would find the idea extremely attractive if it were feasible, but I have grave doubts about this model and its consequences worry me. Your supervisor, JJ Thomson, the man of the electron, had another idea and I wonder why you have given it up. I have heard you no longer agree with him since the work you did in Montréal and your latest experiments in Manchester. Your atom is now very different from the one JJ proposed. I must confess I don't like either of them. As a mathematician, I find the whole story rather puzzling. I don't claim to understand everything you say but I do see some real problems in both of your schemes.'

'You are far from the only one,' answered Rutherford with a laugh. 'All the fire is now concentrated on my atom, but the old "currant bun" of JJ Thomson no longer interests anyone. I guess that is a good sign for me. But tell me: what is it that bothers you so much as a mathematician? Honestly, mathematics doesn't have much to do with it. I am interested in your opinion, but I must warn you. Objections from theorists don't cut much ice with me. Anyway, keep your answer simple. Most of your colleagues fly up into the air with splendid stories which always seem vastly superior to reality. I am only a humble physicist. I only believe what is confirmed experimentally by nature. That is the ultimate test. In fact, there is no other.'

'Perhaps,' answered Schott. 'However, I imagine you would not cast doubt on Maxwell's theory of electromagnetism. It fits your criteria anyway as it is supported by many experiments, performed by none other than Faraday.'

'Obviously not,' said Rutherford. 'Of the two, I must admit I prefer Faraday. Both are great geniuses, but Faraday managed without equations, which I like better. Mathematics does have its limitations.'

'Nonetheless, you do need Maxwell's theory,' answered Schott, 'because it connects electricity with magnetism in a very subtle way. There would be no electromagnetism without it. Because of his theory, I find it very hard to imagine the motion of your so-called "Planetary Atom". I was already having difficulty with JJ Thomson's electrons stuck in a bun. He was claiming them to be point-like particles, which sounds very simple, but in fact is not as soon as you ask what it can be made of and what holds even an electron together. You even want electrons moving around inside atoms. We would have them everywhere. Thomson describes a positively charged cake with the electrons inside. He sometimes referred to the atom as his "plum pudding". I can't help finding this kind of physics a bit strange. Now, you are suggesting in your latest paper something quite different: a very light electron on a permanent orbit around a heavy, positively charged nucleus at the centre. This is yet another disturbing idea...'

'It isn't just an idea!' interrupted Rutherford. 'What you are describing is the result of an experiment. I don't come forward with ideas just for fun.

I rely on actual observations. With Thomson's model, my measurements would make no sense at all. I was observing collisions. To be plain, my experiment was like firing a fifteen-inch shell at a piece of tissue paper and discovering that it rebounds. My projectiles were alpha particles and these are heavy. For them to bounce back, there must obviously be something heavy at the centre of the atom. There is no other way. So, whatever you may say about my model, I am dead sure it is correct, with or without Maxwell.'

'Probably, it will have to be without,' muttered Schott.

'Thomson's model just doesn't work,' added Rutherford. 'All I am saying is that you have to concentrate the positive charge into a heavy ball at the centre and let the electrons orbit around, just like the planets around the sun. If you do that, the experimental results make sense. I can't see any other way. Try if you can to find a better model, but remember that you need to explain my observations, so it won't be easy.'

'It isn't the first time I have come across a hypothesis like yours,' added Schott, who seemed to take little notice of Rutherford's remarks on the importance of experiment. 'I have also read a paper by a Japanese fellow. He has a funny name for a physicist: Hantaro Nagaoka. He was proposing something rather similar. I wrote to him with a few objections, but never received any reply. Maybe my letter got lost in the post somewhere between us and Japan.'

'Really?' said Rutherford, a bit irritated by Schott's attitude towards his experiment. 'I never realized you were so interested in modelling the atom. Actually, if you had read my paper carefully, you would have noticed that I also quoted this Japanese fellow. He was claiming to explain by some kind of astronomical analogy why atoms are stable, but the argument is rather contorted and I don't accept his reasoning. He believes that electrons organize themselves around the nucleus like the rings of satellites around Saturn. Why Saturn? He never explained that. His idea was that the rings are stable. That was all. He seemed to forget that electrons carry charge, so they could hardly arrange themselves into stable rings. They would just fly apart because of the repulsion between them. His argument didn't go any further than just some kind of analogy. Anyway, I heard recently that Nagaoka has given up this model. Maybe that is why you never received an answer to your letter.'

'That was only one of my objections,' said Schott, 'but if you say he has given up his own model, then that is the end of the matter. There is no need to take the subject any further.'

'True enough,' said Rutherford, still a bit annoyed that Schott seemed willing to accept ideas without referring to experiment. 'Anyway, the only worthwhile objections to Nagaoka's model would have been based on observation. Theories are just theories until they are confirmed, and models are just the same.'

'I suppose you are right,' said Schott. 'I sometimes wonder about what is real. One can imagine that some models would just be too difficult to verify. One would then have to find something more than mere experimental tests. Maybe some principle of coherence would come into play. Anyway, since your model is very different from Nagaoka's, my objections would have to be rather different and I need to think harder about that.'

Their scientific exchange went no further on that day.

They began to talk about less serious matters and to share other memories of university life. Rutherford, like many New Zealanders, was a passionate rugby fan and an accomplished sportsman. Schott, however, was not in the slightest bit interested in sport. Apart from mathematics, his main subject was chess. Despite the difference in outlook (or perhaps: because of it) the first meeting between Rutherford and Schott marked the

beginning of a long and lasting friendship. It was based, not only on what had been expressed, but also on a hidden complementarity between their characters.

Rutherford understood perfectly well that Schott, on the first encounter, should reject his model, although apparently he did so for deeper reasons he could not completely express. Schott had the reputation of being a great expert on Maxwell's theory and Rutherford, in the back of his mind, had a few worries of his own about how to reconcile his model with the forces of electromagnetism. So, this was perhaps a useful meeting. He had published his ideas about the Planetary Atom very early, despite not being completely sure. He had to publish prematurely, as he was worried about competition. He knew that many other physicists around the world were trying to catch up with him and said so on several occasions. His main concern was about the group in Paris.

'I have to keep going,' Rutherford said, 'as there are always people on my track. I have to publish my work as rapidly as possible. The best sprinters in this road of investigation are Becquerel and the Curies.' Because of them, he felt he didn't have the time to sit down and work out every single little detail of the atomic model. The important thing was to publish the main idea and have it recognized under his name.

On meeting Schott, Rutherford's first impression was probably that he had finally met the man who could resolve the issues which still bothered him, a bit like James Clerk Maxwell had followed Faraday's work step by step and built up a rigorous theory of electromagnetism. The impression was reinforced by the fact that Schott's personality was rather withdrawn. Not only would this distinguished mathematician be able to help him but, in addition, he would not exploit the opportunity to present himself as the originator of the whole idea, as some theorists are prone to do.

The outcome of the meeting was very satisfactory to both of them. They each came out feeling sympathy and admiration for the personality of the other. Schott, being distinguished but shy, had qualities Rutherford did not lay claims to. Rutherford, on the other hand, had a powerful presence, a loud voice and the healthy physique of a strong farmer, a man of toil and labour. So, they would always be different. Schott was quite particular about sartorial style. Rutherford cared so little about his appearance that he might well have seemed short of money judging by the quality of his clothes. There could hardly have been more contrast between them.

On reflection, Schott was very satisfied that he had come to visit Rutherford in Manchester, despite the wind and the rain on that day. It had been most fruitful. In fact, on his return journey, he began to speculate that his decision to come all the way to northern parts he usually avoided had been the right one. Meeting a person of such substance could alter his scientific direction and might well prove an important event in his life.

The very same evening, Rutherford informed his wife Mary that he had just met an extraordinary mathematician, a high flyer by any standards, who could be expected to solve very challenging problems. They were both quite different from each other, but had one important point in common. Both had studied in Cambridge and at the same college: Trinity Hall.

'We must invite him,' Rutherford told his wife. 'He is one of the great minds of our time. Everybody says so. He is surely a person to know.'

'And,' he added after a short pause for thought, 'he is quite the model of truly British behaviour. A man of style and elegance. I admire him very much, despite his interest in pure theory. After all, mathematicians do serve a purpose. We all do what we are best at. I suppose that is pretty normal anyway. I can't imagine him doing an experiment.'

Eileen Rutherford's
Atomic Physics

ileen Rutherford was only 10 when her father proposed the Planetary Atom. Despite her young age, she soon became conscious that something important was happening in the world outside and was actually about to intrude into her own existence. At first, she thought it would be just a temporary disturbance of family life, like a thunderstorm or snow showers, which can disrupt daily habits but clear away after a few days. Soon, she discovered that the matter was more serious than expected and perhaps likely to last a little longer.

As a young child, Eileen was fully familiar with atoms and knew many of them personally. It could hardly have been otherwise, because her father always referred to them as though they were alive. Since they often participated in the conversation, she had developed a subtle knowledge of their characters. She also knew who was important in their social pyramid. Uranium and thorium, for example, were *the* leading figures of the atomic world. They tended to appear more often and make a more lasting impression in the conversation. Later on, she discovered that helium and oxygen also belonged to the atomic elite, though with a lesser rank. At first, she tried to exclude them from the family, but her father insisted that they did belong and, of course, her father was invariably right about such matters. So, eventually, she accepted them in the inner family circle and even developed a more intimate connection with them. Helium, in particular, she rather liked for remaining aloof from all other atoms, although he was occasionally a little haughty.

However, when it came to the Planetary Atom, she could not help feeling things were going too far. Atoms and planets were really different altogether from each other. For a start, atoms were small and planets were big. Her father had already explained this quite carefully to her. So, why were they getting mixed up? Eileen did not accept the intrusion immediately. She suspected that her atomic family was about to be broadened beyond all recognition and would eventually become quite unmanageable. So, her first instinct was to draw a line and say: no further!

However, once again, she had to make a compromise. The Planetary Atom had come to stay and there was no easy way to get rid of him. He now seemed to occupy all her father's thoughts.

Rutherford was not really thinking about education when he talked to his daughter about atoms and about research.

Of course, it would have been better for him if she had been a son, but since he only had a daughter, he had to make do. Anyway, Eileen seemed to understand him, and she was definitely proud and happy to follow everything he said.

So, although Rutherford was quite convinced that science was not a subject for girls, he fell into the habit of telling her some of his innermost scientific thoughts. Of course, there was no other real subject of conversation in the Rutherford household. The Master of the house was often surrounded, even at home, by students, with whom there was a permanent dialogue concerning atoms and molecules of all kinds and even some particles smaller than atoms whose properties were only just beginning to emerge.

Rutherford was so wrapped up in the exploration of this new domain that he could hardly stop thinking about it and the invisible world of the infinitesimally small had become a question of concern to everybody around him at all times of the day.

Rather than attempt to enter Eileen's childhood fantasies, which would have cost him a considerable effort of mind, he found it a lot easier to draw her into the world of atomic physics and his quest for deeper forms of truth. For him, it was a lot easier and more natural to involve her. So, he shared with his daughter many new ideas she could only half understand. In a way, he was continuing to think aloud next to her and she had become a helpful witness of his continuing research. Eileen, on the other hand, was quick in absorbing new words, new expressions, sometimes even complete sentences. She did not really need to appreciate their full significance in order to use them correctly. She knew the context and that was enough for her.

In fact, she had a remarkable vocabulary for a child of her age and expressed herself almost like an adult. Often, her father's ideas would come to the surface, and she seemed to be anticipating future times in what she said. This could create an illusion of deep understanding. Even her father was sometimes surprised at what she would come out with.

That is how Eileen Rutherford became (without even realizing it) one of the first persons in the world to speak with authority about the Planetary Atom, which much amused her mother, Mary.

It must be rather difficult (she had some inkling about that) to be the daughter of a great physicist. If there was one advantage in this situation, it was her familiarity with mysteries other people come to terms with, usually after years of study. Eileen was up to date with the very frontiers of science, from which she received unsolicited daily reports.

In one way or another, daughters need to attract the attention of a father and to feel loved and admired. For Eileen, this was a bit complicated. Rutherford, despite great moral qualities, was usually thinking about something else when they were together. So, it was really up to her to take the first step and be noticed. That is how she learned the trick of completing his sentences whenever he paused, usually so accurately, using his own words to do so, that Rutherford hardly noticed any hiatus. Sometimes, he would suddenly realize that she had spoken and be surprised. Sometimes, he didn't even notice. But, when he reflected on their conversation, he almost had the impression that she had understood the very complex ideas he had put to her. In reality, she used a mixture of love, good memory and guesswork guided only by feminine intuition. Thus, the dialogue between Rutherford and his daughter was rather one-sided. In fact, the scientist was talking most of the time to himself but had an impression of joint progress, as he had to organize his thoughts before presenting them to her.

He would often cite an adage familiar to good teachers: it is by better and better repetition of the same ideas that one ends up understanding them fully. However, his opinion about female intelligence was oddly ambiguous. He often said to his students:

'A good idea is one you can explain in simple terms to a barmaid. If you can't get it across to her, it isn't her fault. There are only two possibilities. Either you didn't explain it properly, or it just wasn't a good idea.'

Nobody ever met Rutherford's barmaid and there is no other trace of this lady in the history of modern science. Most probably, Ernest Rutherford invented her. It is possible, however, that Eileen really played her part. Only, Rutherford did not want to mention his daughter in this way, for perfectly proper reasons of family discretion. Also, since he was talking to young university students, it was generally more appropriate to introduce a barmaid in his argument than to refer to his own daughter.

In fact, Rutherford's 'barmaid' stood in for the whole of the feminine gender. Rutherford was prudent in how he expressed it, but there was clearly an underlying misogyny in his mode of thinking. The subject was already sensitive at the end of the nineteenth century, as one can tell from the care he took to cloak his opinions in a variety of metaphors. In summary, his argument really went as follows:

If I can formulate my discoveries in such a way that even a female mind apprehends my meaning, then I will have achieved something worthwhile.

But he did not dare say things quite so explicitly, hence the mythical 'barmaid' known to all his students. A 'barmaid', all of them would probably conclude, had only half the intelligence of an average barman but, again, this was never actually said.

Despite all Eileen's affection for her father, she was left wanting some kind of return. She had the impression of climbing a tall mountain to reach him, while he remained perched at a great height without ever acknowledging her. Rutherford was able to untangle the hidden secrets of Nature, but not the psychology of those closest to him. That, in a sense, was his greatest failing.

He was fortunate in enjoying domestic bliss, with a doting wife who brought him as much feminine attention as he could possibly need. There was no call for anything more. His life ticked over as regularly as clock-work and his environment in Manchester was the epitome of the research establishment of Victorian England. It was entirely masculine in its outlook. As everyone knows today, there should always be some feminine input, however small, to avoid the rigidity of a stiff society.

At the periphery of Rutherford's professional circle, there was, it is true, just one feminine influence strong enough to reach Eileen: that of the

French researcher Marie Curie. This distant lady, intruding from a foreign land, was something of a mystery to her. Eileen was much troubled by her permanent presence in her father's sphere and tried hard to understand where her power lay.

Like the atoms, she was not to be seen by everybody, but she was definitely in Rutherford's thoughts. In fact, she seemed strongly connected to the most powerful and important of all atoms, one so strangely frightening that it even eclipsed uranium. The French lady had found an atom capable of shining in the dark with a mysterious light her father had been privileged to see with his own eyes. Since then, she had become Madame Radium, or at least her influence appeared to be entirely based on this strange new element. In some way, without even coming to Manchester, Marie Curie could occupy a whole conversation between Rutherford and his students. Clearly, a woman could occasionally play an important part, important enough to equal even the Planetary Atom in the themes of the research team. But how did she manage that? For Eileen, this was difficult to understand. She did not know that 'Marie Curie' when Rutherford mentioned her name, was not thought of as one solitary woman working in Paris. She was a whole team, including her husband Pierre Curie and even old Professor Becquerel, the discoverer of radioactivity, not to mention technicians beavering away in Paris. She could hardly imagine, because it was always Marie Curie who penned letters to Rutherford, that her father was communicating with a powerful international group every time he wrote her a personal letter. Nor did she understand that the physicists in Paris were competitors Rutherford himself had chosen to set himself up against.

The importance of atoms was obvious enough to Eileen. The importance of Marie Curie was rather less intelligible to her.

The problem was that Marie Curie was a person. There was even a photograph of her in her father's office — a troubling image of another lady, which even carried some incomprehensible words scrawled above the signature. A real woman was rather difficult to transform into an absent family member, as Eileen had successfully achieved with atoms. There was a need for more explanation.

Usually, Marie Curie did things without consulting her father. She sometimes wrote expressing views he strongly disapproved of. How could this be allowed? Madame Curie was clearly a problem, not only for Eileen but also for Ernest Rutherford himself. She was clearly out of line and could hardly be considered on a par with uranium as a loyal member of the extended family.

So, there was no denying it, Madame Curie was a kind of strange exception to be reckoned with. There was surely no lady of this kind amongst those Eileen had met at home. The family friends of the Rutherfords did not include female physicists of any kind. Indeed, it was hard to imagine such a peculiar person in England. Her father had explained that, in fact, Madame Curie was not even French. She actually came from Poland, a country rather close to Russia. These two places were even more distant than France. They shared not a single frontier, not one custom, nor the slightest tradition with the British Isles. The tentative explanations provided by her father made things slightly more plausible but fell far short of accounting for so much strangeness. Eileen knew nothing much about any of these countries, except perhaps their rough positions on a terrestrial globe. However, she felt that rules of behaviour are rules of good sense and must apply universally, even in Poland.

This lady claimed to be a physicist like her father. This was already surprising and incongruous. She also, apparently, had forced her way into an area of knowledge into which her father had introduced his loving daughter, but she had no special right of entry and, more importantly, was presumptuous enough to disagree with him.

There was something odd about this French or Polish lady and Eileen was fascinated by her biography. Apparently, there was much more to be discovered than her father was willing to reveal. Since many details were missing, Eileen decided to fill them in somehow, as she had done so successfully for uranium and thorium.

Maybe, she decided one day, *this lady has run away from Poland because, even in that country the idea of lady physicists is unacceptable to honest people. My father never dared to tell me about that for a very simple reason: he also realizes that her behaviour is really quite shocking. Since science isn't appropriate for girls, she had to emigrate to a foreign country to hide what she was doing. Even there, she practises physics secretly because, even in France, it can't be quite the done thing. That was clever of her. There isn't much the French can do about it because she is in fact Polish. How did she ever think of that? Since nobody knows her, she can do anything she wants, even become a physicist.*

As a country for an ambitious lady to run away to, France seemed a good choice to Eileen. She knew something about it, because France is the country most talked about in England, providing all sorts of good examples of bad behaviour. She had worked out that, for the French, transgression was a way of life, almost an established rule. Marie Curie was obviously a revolutionary, a Republican. She clearly belonged to a dangerous species people sometimes referred to: women who do whatever they want, including of course rather bad things nobody dared mention. Of course, this was quite tempting to imagine in conversation, but such excessive freedom, she knew full well, was not on offer. Nor was this the conception of femininity the Rutherford *milieu* would find in any way acceptable.

However, there was a paradox. Despite the disapproval which should have been expressed, nothing was said explicitly against Madame Curie. When her father received a letter from Paris bearing her name on the envelope, he would actually put everything else down in order to read it. This seemed inconsequential to Eileen. If he disapproved of women studying science, why did he make an exception only for her?

In fact, although she did not admit it to herself, Eileen could not help feeling this was unfair. After all, she did more on a daily basis to support her father's highest flights of scientific thought than a lady living miles away and in another country, who just wrote him a letter from time to time. Her father seemed to think encouragement was completely normal coming from his daughter. He could at least have acknowledged how remarkable she was in following him in his footsteps, but no. He seemed to think that a letter from this Polish lady was somehow more important. Eileen was rather unhappy about the influence of Marie Curie and would have preferred a planet without such an unnecessary country as Poland.

Rutherford liked talking about science to his daughter but did almost nothing to help her understand it. She was left to her own devices to make sense of what he said. Although Eileen was fascinated by the strange world of research, there was no helping hand to open doors before her. He was famous for being direct in all he said and telling everybody about his new ideas, but the great joys of discovery, such as seeing for the first time something nobody had ever seen before, were emotions reserved for a small circle of students. Even though he talked fairly freely about his work, the reality of the laboratory was only accessible to a very privileged group around him and certainly not to a small girl.

To Eileen, this was a sensitive issue. She did not really understand why she was not allowed into this inner circle. She came to the conclusion that, despite all her efforts, there must be some respect in which she did not quite make the grade in her father's eyes. Perhaps he did not want to tell her so. It was obviously rather difficult to be his daughter. Maybe she was not clever enough. Or else, it was something to do with not being the boy her father would have preferred to have. It was a rather humiliating thought, which she kept from her mother. What could she do about it anyway? She couldn't help feeling that she had done her best to rise to the challenge. Why did she feel a sense of guilt about being who she was?

Once or twice, when she found exactly the right words to complete a sentence Rutherford had begun, she caught an admiring look from her father which cheered her up a lot. A few moments later, however, she was filled with doubt again. It was an unpleasant feeling which gnawed away at her and seemed impossible to share with anyone.

One day, she overheard a conversation between her parents which interested her a lot. It was about diseases which affect the mind. An Austrian doctor claimed that he could do something to treat them. It was, said her father, perhaps the dawn of a new science, but the question was: is this real science or not? He took the view that it was too early to say, because it seemed difficult to achieve experimental confirmation. Eileen would have liked to help this doctor prove he was right. Suddenly, she realized that she probably had symptoms of such a disease, though, she hoped, in a mild form. She would have loved to find out more, to learn something about this new science, maybe even to meet the Austrian doctor, but of course, that was just fantasy. There was no possibility that such a meeting could ever happen. Anyway, the thoughts of Rutherford's daughter could never become a subject of scrutiny. The mere suggestion that they were sometimes unhappy would have been impossible to formulate. In the Rutherford household, there was one and only one religion: effort and self-control leading to triumphant success. Any other outcome was inconceivable.

So, Eileen regularly returned to the task she had set herself of trying very hard to understand what her father was really talking about. Despite her young age, she had made quite a lot of progress in this direction. For example, she had worked out that the electron was like a tiny marble carrying a negative charge. There was a lot of talk about electrons in the Rutherford household. The reason was that her father had been a student of the great Professor Thompson in Cambridge, usually referred to as 'JJ' by people in the know — the discoverer of the electron and of all its astonishing properties.

She also knew that the very same JJ had proposed a model for the atom. Why an atom needed to follow a model in the first place was a bit beyond her understanding, but if it really needed one, then the currant bun was definitely her favourite, with little electrons dotted inside like the fruit in a cake of positive charge. All of this was pretty clear to her and she followed in every detail the explanations given by her father. In fact, she had become something of an expert on atomic pastry.

So, she was understandably disappointed when Rutherford began to destroy this simple and homely picture in order to replace it by the famous Planetary Atom which was now all the rage. He explained it to her using an astronomical metaphor, with electrons circling round a heavy, positively charged nucleus, just like planets about the sun. Eileen was not too keen on this new idea, but she realized that the physical world, far from being fixed and permanent, could suddenly change, a bit like the weather, when something new was discovered. That was the whole point of physics. So, she didn't complain too much about it. She could well imagine that everything might be upset again a few months later when some new discovery was made. The planetary hypothesis would rule for a while, but would surely be replaced by something else which might prove more agreeable. In the meantime, she decided it was better to go along with the idea, which was anyway necessary to maintain her position close to her father. Perhaps, one day, she might think of something new herself. At present, she felt her own mind was quite barren. She knew that he only admired scientists with formidable powers of imagination who could create a new vision of the world, or upset all prior notions in some radical way but, of course, that was asking too much of a young girl like her.

She had reached this point in her perceptive self-analysis when Rutherford introduced his family to someone very different from the young research students she had met so far, a man just slightly older than himself, for whom he professed a great admiration. His name was George Schott. From their very first meeting, Eileen fell under the spell of his new and unexpected personality and was attracted to him as she had never been previously to any other person.

He was just different from all the others. At first, she was staggered and even rather shocked to observe that Mr. Schott, whose name had never even been mentioned to her before, spoke to Rutherford as though he were an equal. His tone was quite independent, without any of the deference students generally adopt towards a respected teacher. This was different from the usual style in Manchester, where Rutherford was the most prominent member of the Faculty. Apparently, this was not just a matter of age. Schott came from another world, a distinct intellectual sphere, whose importance was at least comparable to the University of Manchester. In this alternative world, the stars were not quite the same and they shone in a different way. Schott could follow paths which were not identical to those of the Rutherford school and yet the Master accepted them and was open to dialogue. When he introduced this new visitor to his wife, he just said:

'This is Mr. Schott, one of the great mathematicians in the land.'

Even this introduction was a bit surprising to Eileen. Usually, her father would only praise experiment and make somewhat disparaging remarks about theory. But Schott was there, standing before her, and quite clearly her father had a high respect for this newcomer. How could this be? How had this new state of affairs come about? In the conversation which followed, Schott expressed bold views completely at loggerheads with those of her father, and yet Rutherford accepted the contradiction with good humour, even attempting to state his opposing opinion as politely as possible. This was a revelation to Eileen. She had never witnessed anything like it.

As the evening progressed, she also began to notice many small details which made Schott's personality quite extraordinary in her eyes. There was an air of mystery about him, the air of a person who knows a lot more than he cares to declare. He avoided simple arguments. He called them 'pictures' and seemed convinced they could never be quite satisfactory. He regarded them as incomplete.

Another strange characteristic was his aversion for sport. That was very unusual in Manchester.

Schott seemed to hate anything rough and had a highly developed taste for subtle reasoning under all circumstances. Coming from a man, this was very new to her.

Then, he was clearly concerned about his own appearance and his clothes were extremely refined. This was something Eileen had not encountered before. By comparison, her father dressed in a drab and rather ordinary way. This man Schott clearly moved in different circles, where sartorial 'chic' was probably much more important. Eileen decided that, in her own small way, she would try to influence her father's taste in future and improve his style.

There was something naturally aristocratic about George Schott which struck her imagination. She had never seen a man display elegance.

In her experience until then, only women were allowed to concern themselves with their own appearance. Respectability, honesty and righteousness — eminently virile qualities — were somehow all connected with the distinction between the sexes. Men were not supposed to reveal any complexity of taste. They were only expected to be forthright.

'I am a simple person,' was one of Rutherford's favourite phrases, used to good effect whenever a student embarked on some complicated mathematical reasoning: 'tell me the story otherwise.'

Abstraction, on the other hand, appeared to be Schott's preferred mode of thought. He seemed to take pleasure in subtle developments, and always added what he considered essential detail in order to make an argument aesthetically perfect. Eileen was surprised, when observing the outcome, to discover that he was usually right. What she had thought simple was only superficially so. What he had added turned out to be important. After he had pointed it out, others also found it impossible to think in another way.

From their very first meeting, she felt strongly attracted towards the unusual personality of Mr. Schott and even tried to explain the feeling to her mother. Much to her surprise, Mary Rutherford understood perfectly well what her daughter was trying to say and even added words of her own to explain why her husband was so impressed with him.

'They both studied at Trinity,' Mary commented, 'and Mr Schott was the top student in Cambridge several years earlier than your father. Some people say he is the best mathematician in England, and that may well be true.'

In other words, despite her father's disparaging remarks about theory, good mathematicians did deserve some kind of respect, and it was even necessary to listen to them from time to time. Eileen took stock of this new reality she had never suspected.

On another occasion, when Mr. Schott returned for a visit at the Rutherfords' house, her mother pushed Eileen forward towards him and, to her intense embarrassment, informed their guest with a laugh that he had conquered another heart and now had a new young admirer within the family. Eileen was deeply ashamed by her mother's words and blushed harder than she could ever remember. She was so embarrassed she couldn't think what to say. She should really have been annoyed with her mother for such lack of discretion, but at the same time, she felt almost relieved. Something had come out into the open she would never have dared to express herself.

Mr. Schott was almost as embarrassed as she was by the incident. But what he did was very gentlemanly: he spoke to her with such politeness that she felt treated not like a young girl, but quite like a lady. She was both flattered and somehow rather happy with the outcome. Above all, she was grateful that he had not taken advantage of the situation to say anything which might have embarrassed her yet further.

George Schott returned several times to visit the Rutherfords. There were a number of important points of physics he wanted to understand about the famous 'backscattering' experiment. At least, that was always the official reason for which he returned. However, every time, he would spend as long as possible in conversation with Eileen and always spoke to her with great kindness.

There was a kind of secret complicity between them. She was so clever at following the thoughts of the interesting stranger her father had introduced into the house that she began to discover some curious psychological problems behind his charming ways. One, in particular, went close to her heart because it was a feeling she knew well. Far from possessing the trenchant character of her father, Schott was often full of doubt about himself and his own abilities. He confided that he often suffered from depression, just like her. The only cure he had found for his ills was to seek refuge in the complexity of higher mathematics. Far from being a sign of inner strength, his ability was in fact born of uncertainty about himself. She found this confession highly appeasing and so, it seemed, did he. Maybe it was easier for him to explain this to a young girl than it would have been to make the admission to a grown woman.

There was something about George Schott which made Eileen realize that they were both extremely similar in character.

Perhaps he would have been interested like her in meeting the Austrian doctor she had heard about. In fact, although she previously had the impression her thoughts were sometimes too complicated to be unravelled, George Schott seemed to have reached an even higher level of intricacy.

He told Eileen a lot about himself and even revealed (under seal of secrecy) that he was originally not entirely British but secretly German by family and by culture, so that he sometimes felt ill at ease in his country of birth. At least half of his thinking came from elsewhere, from a fatherland his parents had chosen to leave behind for reasons he was not completely clear about himself. Despite this, the Schotts remained strongly attached to their traditions, and his family always spoke German among themselves. His closest cousin, who lived in Frankfurt, was like a second brother to him and often came to stay with them in England.

After so much intimate conversation with George Schott, Eileen felt
they had become close friends and, one day, was emboldened to address
him by his secret name Adolphus, the name only his family knew him
by, which she pronounced very delicately and with great tenderness. He
understood why at once and was very touched. She had guessed that this
might become a discrete but telling sign of their friendship. Others around
them might overhear the name, but would not really know what to make
of it, even if it did allude to some sort of secret pact between them.

Chapter 4

Why Ernest Rutherford Needed George Schott

The contrast between Ernest Rutherford and George Schott, both physical and intellectual, could hardly have been greater. Rutherford, brought up as a farmer's son in the countryside, was the healthiest of men. He was also very direct and open in everything he said. In his household, there was, as one would say today, complete transparency. Nothing was ever done which might seem better kept secret.

Rutherford's voice was a very loud one and the very sound of it somehow denoted the great sportsman he was. There was a powerful physique behind, as well as a strong personality, which marked him out as a natural team leader.

The Schott family were a very different kind of people. They certainly belonged to a prosperous middle class but, even so, lacked the confidence people usually associate with wealth. In fact, they all had hidden layers of character which were hard to disentangle. There was always something ambiguous about their behaviour. They seemed unsure of everything, including themselves, and took care to keep out of the limelight, as though something about them needed to be hidden. George Adolphus Schott had inherited much of this odd behaviour. Members of his family hesitated between countries. They superposed several different cultures, suffered from strange ailments and their health was a permanent matter of concern.

Another family peculiarity was the large number of 'Georges' in the house. First names were attributed through a mixture of German and British traditions, but repetitions and redundancies revealed a lack of parental imagination. If anybody had stepped into their house and called out: 'George!' several people would have appeared simultaneously. Owing to a bizarre family tradition 'George' had become the first name of all Schott's of English descent, complemented by a more Germanic second name such as 'Gustavus', 'Adolphus' or even 'Augustus', always with a Latin ending, which was very fashionable at the time in German families.

More revealing, perhaps, of a certain snobbery amongst them, was the systematic choice of first names with monarchist undertones. For example, Gustavus Adolphus was a clear reference to the famous king of Sweden. The Schotts, like many prosperous middle-class families, had unfulfilled social ambitions which betrayed themselves in their choices of heroes. In fact, the parents of George Adolphus Schott were simply German immigrants. It was not quite clear what had brought them to England. Ostensibly, Mr. Julius Schott (the father) who was an engineer had heard about a technological revolution taking place in Britain. He wanted a share of the action, so he brought his young wife with him and they settled in West Yorkshire for reasons connected with industry and manufacturing. It was a good move for an enterprising and talented young engineer. Julius Schott was soon involved in the design of new types of heavy machinery to exploit the power of steam. Mrs. Amalia Schott was also an enterprising person, and she joined in the social activities of her husband with the firm intention of playing her full part in the family success story. She had a high idea of her own worth as well as that of her husband and was by no means second to him in ambition. On the birth of their children, she insisted on choosing the most distinguished forenames she could think of, and this was a subject of sometimes tense discussions with her own husband.

Julius Schott was somewhat more discrete than his wife, because he had more direct experience of the workplace. As a foreigner, he was aware that a bad choice of names could subsequently draw unpleasant comments for a whole lifetime, so he was inclined to err towards caution. The Schott family had brought with them from abroad many of the prejudices of the world they came from. Although they belonged to a new caste of inventors and innovators, whose achievements were based on cleverness and hard work, they lived outside British customs and a suitable choice of names for their children had been a bit of a conundrum.

Madame Schott was interested in history and fascinated by the great military successes of king Gustavus Adolphus 'the Great' renowned as the 'Lion of the North' in Germany. She had therefore decided quite early in life that her two first sons would be named after him, as 'Gustavus' and 'Adolphus'. Such prestigious names would surely serve them in good stead during their careers.

Her husband strongly disagreed with this idea. He was just as ambitious as his wife, but much more realistic. He feared that such foreign-sounding names would only attract ridicule and that the connection to a king of Sweden, if noticed, would only make matters worse. He did not want his sons teased in the workplace as he, Julius, had been.

He was worried about latent xenophobia and was sure such names would be a very bad choice. So, they quarrelled a lot about the subject without finding a good compromise. Madame Schott had a strong argument: the money needed to set up a small engineering company they both wanted to launch came from her side of the family, which was well-heeled. Clearly, Julius Schott could not dismiss his wife's wishes out of hand. But he did argue the point with her and tried to make her understand that such German names alone would not really be a good idea and should at least be accompanied by something truly British.

The problem was: how to select a British name? Julius Schott, deep down, was just as vain as his wife. He just wanted to be more subtle. It was actually by a rather complex argument that he finally hit on the name: George. It is really a very ordinary name. At one time, it was common to address all taxi drivers in London as 'George' — a kind of snobbish familiarity which fortunately has died out. Julius Schott was of course unaware of any such connotation, and had a completely different reason for choosing this name. For him, it was a way of combining German with British history. He was himself originally from Hanover. *Diplom-Ingenieur* Julius Schott spoke English with a strong German accent and knew that he could not hide it. Indeed, it had been the subject of many jokes by his colleagues. So, he found a brilliant way of reminding his friends, relatives and employees that George the First, Elector of Hanover, had been selected by the British Parliament as the new king of England after the end of the Stuart dynasty. The Elector had travelled all the way, by carriage and by boat, and arrived in London in the company of two women (one his legitimate wife, the other not) one tall and thin, the other short and fat, much to the delight of all the cartoonists of the period. He had come to rule the nation and was very proud to have been selected by the British Parliament as the first and foremost Protestant Prince with a valid connection to the House of Stuart via the Electress Sophia.

What a wonderful example! thought Julius Schott. *After all, King George never spoke English any better than I do. This way, I show respect for my adopted land and also remind the English of their strong ties to Germany through the royal family. Socially, it is perfect. What could be more distinguished for our son than a name borne by English monarchs?*

So, the compromise was obvious. The eldest son of Julius and Amalia Schott was christened George Adolphus. When the second one came, it turned out that it would have been extremely difficult to do any better for the forename, so Julius Schott was rather stuck until he realized that the second name would suffice to distinguish between them, so he became George Gustavus.

It must be said that the strange mix of a very ordinary British name with a pretty unusual and warlike Swedish name seemed somewhat clumsy and a difficult choice for the children. Julius Schott was a fine engineer, but somewhat lacking in imagination when it came to christenings.

From their earliest age, the two children had to live with the difficult choice their parents had made for them. Both of them were always embarrassed when obliged to reveal their second names, which they could hardly avoid at school, since there was no other way of distinguishing between them in class lists. They were rather more lucid than their parents regarding social ambitions. They knew instinctively that their family prejudices and customs bore little relation to the British social fabric. As immigrants and foreigners, they had very little chance to rise any higher than their existing status — the professional middle-class — in the rigid hierarchy of Victorian society.

George Adolphus was particularly troubled by the unfortunate inheritance of so much family snobbery. He could not say much about it, but it became second nature to him to avoid any mention of a German family connection if at all possible. He was so convinced that he would rise no further in British society that he became spontaneously self-effacing and took pains to avoid any prominent role.

Luckily for him, such discretion is regarded as a social asset in England and much appreciated. Nothing could be more vulgar in English society than the American habit of trumpeting one's own achievements.

Both of the Schott children, as it turned out, were highly gifted at mathematics, and Madame Schott was convinced that they could build on such talents to rise to the highest levels of society. This, however, was quite unrealistic on her part. The upper class, in England, often called 'the Establishment' is a pretty inaccessible pinnacle, open occasionally to persons of extreme wealth, but not generally to foreigners and certainly not through the study of sciences as might be the case in Germany. It remains basically an aristocratic class even today and was even more so at the end of the nineteenth century. The British aristocracy lays claim to a long history and has little time for modernism of any kind. Engineers like those in the Schott family built their fortune and success on technical progress and innovation. Had they followed the path imagined by their parents, the two brothers would have been trapped in a dead-end. Luckily, they both saw the peril in good time and did not display such unrealistic ambitions.

The great merits of Julius Schott were his tenacity and commitment. Thanks to him, the family company went from strength to strength. He had set it up in Bradford, where it fitted in perfectly with the local pattern of activity. The town was going through a period of intense industrial development, and the Schott family participated with all their might in the transformation of its infrastructure and in the Industrial Revolution which was creating the foremost modern economic power in the world. Obviously, there was a lot of self-interest in joining this surge forward, but self-interest could very well be construed as a virtue.

'We are bringing prosperity and happiness to the people of Yorkshire as well as to the country,' said Julius Schott on several occasions. 'Thanks to us, the inhabitants of Bradford find jobs. In return, it is normal that we should reap some benefits. We also deserve our share.'

Neither Amalia Schott nor their children had any doubts about this philosophy. However, the activities of the factory, although they undoubtedly contributed to the wealth of the nation, also brought some quite serious problems, not only for them, but also for the little town where such businesses had mushroomed. The Schott and Sons Weaving Company was far from being the only one in Bradford, but played its part in polluting the whole area.

Today, Bradford, where the Brontë sisters lived for many years, is a clean and charming town. Their amazing literary reputation so influences the imagination that one immediately pictures a romantic environment worthy of *Wuthering Heights* or *Jane Eyre*.

However, the reality at the time was very different from what it is today and, of course, bore no relation either to the famous literary works. It was, in fact, entirely tragic. Bradford in the nineteenth century was more like an industrial prison. This may actually explain many aspects of the Brontë sisters' short lives. Often, one thinks of them as victims of the great epidemic of tuberculosis which caused such heavy mortality in the Romantic period. Chopin, Keats and other famous artists come to mind. However, that was not the reality for the inhabitants of Bradford. They suffered, indeed, from respiratory diseases, but the real cause was not contagious tuberculosis. The disease was another. Through its industry, Bradford was rapidly becoming the most heavily polluted city in the world. Without realizing it, Julius Schott had brought his own family into one of the most dangerous places on earth. And, of course, his own factory was contributing to the problem. The engineering pioneers had no idea of the hazards. Such conditions had never been created before. Initially, nobody had any inkling of the consequences.

The signs, in fact, were obvious from the start, but their interpretation was not clear to those who created the first industrial estates. When Amalia Schott put on white lace gloves to go to church on Sundays, she became used to finding them quite black on her return. A few hours were enough. She attributed the effect to the climate in West Yorkshire, and thought no further about it. She was neither a chemist nor a scientist of any kind.

Bradford and its surroundings, at the time, were simply drenched in industrial pollution. Julius Schott, as an engineer and factory owner, was actually one of the first to suspect a health problem and to have some doubts about burning coal. He noticed that employees in his factory were dying at a particularly young age. The average life expectancy in Bradford at the time was eighteen. He attributed this sad fact to the hard lives manual workers led, but was concerned there might be some other cause, and began studying mortality figures in the region. He even attempted to involve other factory owners in his investigations but was unable to interest them.

What he had noticed was true. However, there were several related causes. In addition to the direct pollution of air by smoke, there were secondary effects. Because of the anomalously large number of deaths in the town, the underground water was contaminated and there were regular out-breaks of typhus and even cholera. Other factory owners seized on this as the real explanation and argued that high mortality had nothing to do with smoke or air pollution caused by factories. They also incriminated alcohol consumption, as well as the abuse of laudanum (a mixture of opium and alcohol quite freely on sale in Yorkshire, which was often used as medication).

The truth is that the prime cause was polluted air. The workmen in the Schott factory, Julius Schott himself and his young family were all breath-ing it, day in, day out. It was not just grey or black but even, at times, yellow, and everything which came into contact with it eventually turned yellow as well. The worst was the permanent smoke and fog in front of the furnaces where the coal was actually burned. The chemical tuberculosis which was poisoning the workers' lungs was due to the high percentage of sulfur in the coal mined near Bradford. Middle-class families lived a little further away from factory chimneys and, of course, were never directly involved in shovelling coal, so they suffered less, but nobody could escape completely. It was simply dangerous to live in Bradford.

When children started to develop symptoms in a well-off family, one would consult a local doctor and he would go through a ritual, muttering

something about consumption then nodding rather sadly when asked about treatment. Julius Schott was a humane and dynamic manager. He believed in science and called his local doctor to the factory several times to examine his employees. As an engineer, he was fundamentally convinced that anything broken can be repaired, including the human frame, so he was unimpressed by the responses he received from practitioners in Bradford.

Through professional contacts, he was able to consult a Harley Street specialist and spent a considerable sum of money to bring him up all the way from London to visit the Schott and Sons Weaving Company.

Even before reaching Bradford, Professor Joseph Brown, MD, was well aware of what to expect. He knew of the town's unenviable reputation as the most polluted conurbation in the kingdom. The situation was well-known to specialists, but the doctors in those times were accustomed to providing reassurance rather than facts, especially on a question bound to affect the national economy very directly. It was important not to spread alarm about such matters in the general population.

The train he had boarded in London soon left the elegant suburbs and green gardens surrounding the town, then entered the grey and dull countryside beyond as it headed towards Manchester. It was, of course, not too difficult to surmise what was troubling some factory owners in the manufacturing areas of the country. As the train made its way towards Bradford, the omens became more sinister. Since 1880, the weaving and dying industry of the town had become the largest centre of its kind in the whole of Europe. According to what he had read, over one thousand workers were employed there. For such a small city, this was a huge activity, dominating the skyline all the way along the railway track. In just a few years around 1880, more than two hundred high chimneys had appeared in the town, spitting out heavy black smoke all over the region, driven one way or the other according to the prevailing winds. The surrounding countryside also was obviously affected. Yorkshire coal contained a high percentage of sulfur and, as a professor of medicine, Joseph Brown knew enough chemistry to conclude that the air in the city must be strongly acidic. He had always heard of course, that Bradford was a mining town, but even he was surprised to discover quite how much locally produced coal was being burned there. The phenomenon was impressive in scale. The Schott family, like all their competitors, had set up huge machines for large scale combing and weaving of cotton and wool. They were powered by enormous steam-driven pistons. The activity had grown at an incredible rate and was now the main economic resource of the whole district. Even towards noon, when he arrived in the centre of Bradford, Joseph Brown noticed that daylight was dimmed. The sun had disappeared behind a thick blanket of fog. There was an eerie feeling of darkness in the middle of the day, and the doctor had no difficulty in understanding why Engineer Schott had been so insistent he should visit the family factory. Indeed, when he had heard that Bradford was heavily polluted, he had imagined some degree of smog hanging in the air, but pollution on this scale was something new even to him.

When he entered the precincts of the Schott and Sons Weaving Company, he was greeted by a foreman in front of quite an amazing sight. Huge black fly-wheels with thick metal spokes, driven by outsize pistons, provided the energy necessary to the company's many and diverse machines. They seemed to rise straight out of the ground of an enormous warehouse, projecting great plumes of steam into the air. There was a constant clanging of metal against metal and an acrid smell hung around them.

At first, he thought this would last only a moment, so he waited patiently for the noise to stop. But after a while, he realized this was not going to happen. This level of noise was the permanent reality of life inside the Schott Company. Neither the pistons nor the wheels ever stopped. New plumes of steam appeared as the motion continued and more wheels began to move. Perhaps it would turn even noisier in a few moments. So, Joseph Brown decided to fight the din just by shouting loudly:

'Is it always as bad as this?' he enquired of the foreman. 'Don't these wheels ever stop going round?'

'No Sir, they never stop,' came the proud answer. 'The wheels of our factory work day and night as we do, week after week, winter and summer. That is what it takes for us to beat the competition.'

'What? Do you even work on Sundays?' asked the doctor. 'Do your workmen agree to that? Even on the Lord's day?'

'Obviously not,' said the foreman. 'We have another team for Sunday work. But I don't know that bunch. They have their own rules and most people don't like them too much. We call them the "desperadoes". They live like pirates and don't really do proper work. Just keep the furnaces going, so that the fires never stop. There isn't any weaving on Sundays.'

'So, the smoke lasts all week does it? Don't you ever stop the furnaces?'

'No, we never stop those,' confirmed the foreman. 'It wouldn't make sense. Stopping and starting furnaces would be a huge waste of time and money. The men on the Sunday shift are a pretty rum crew anyway. They wouldn't find any real work elsewhere. So, they have to be here all day Sundays and even overnight. By keeping the furnaces alight, we can start the factory up again early on Monday morning and no time's lost.'

The London doctor was used to the relaxed way of life of a prosperous capital city, and the frenetic pace of industrial production in Yorkshire staggered him. He had never witnessed at close quarters the conditions under which Britain's prosperity was achieved. He noticed the thin layer of soot on all the objects in the factory, the speckles of black dust on the foreman's face, his hollow voice and bloodshot eyes. As a doctor, he knew what to make of such signs.

'Take me to the Manager's office,' he said simply. 'I know he is expecting me.'

Indeed, Julius Schott was waiting for him and not a little anxious to hear the opinion of the great professor of medicine from London. The results of the consultation were important to him in many respects.

'I think I should begin by examining you,' announced Professor Brown, 'because I presume this is your normal place of work. I am sure you spend as many hours in your factory as do your workers and there is no doubt such an environment presents risks for the health. In many ways, they are actually quite comparable for all of you.'

'So, you do believe smoke is the problem?' asked Julius Schott anxiously while the doctor tapped away at his chest.

'It is not normal in our profession,' replied the doctor, 'to discuss worrying information with a patient. However, you are a man of intelligence and it seems that you have already studied the matter a fair amount, so I cannot hide the truth from you. By breathing in smoke and fumes carried by the air of Bradford, you have not improved the state of your lungs. It is probably much worse for the workers in all the factories here, since they live all day in the dust and smoke of the furnaces. I would guess that what is left of their lungs is in a very poor state indeed.'

'What can be done?' asked Julius Schott.

'The problem,' answered Joseph Brown, 'is that exposure to the air in Bradford is permanent for everybody living here. The only real answer would be to stop the factories from time to time...'

'Stop the factories?' exclaimed Julius Schott. 'That is absolutely impossible! Even a short interruption in our production would be quite ruinous...'

'If your family, and especially your children, are also living near here,' advised the doctor, 'then I think I should see them as well. The smoke fumes in your town are obviously very toxic and the consequences for the respiratory tract of children are far more severe than for adults. So, while I am here, I suggest you take me round to your home. I will be able to examine them all and give you an opinion.'

'Don't frighten my wife too much,' said Schott. 'She is of a nervous disposition, but I will take you to see my family.'

Julius Schott had suspected it for some time, but was loath to admit it, even to himself. The whole of his family had chest problems. They had all been coughing, wheezing, spitting and complaining that it was hard to breathe without ever realizing what the real cause was. Now that a proper specialist doctor was available at last, it was high time to have them all examined and establish a serious diagnosis.

Professor Brown soon confirmed the situation the engineer had begun to suspect. They were all was suffering from the same affliction to varying degrees. For him, distinguishing between contagious consumption and lung disease caused by breathing in acidic fumes presented no problem. He understood completely the dilemma of a factory owner, whose livelihood depended on keeping all his machines running. The Schott family, like all the others in Bradford, had no real choice.

At the end of the day, there was not much he could suggest but, before leaving, he tried to come up with some useful advice, especially as concerned the children.

'I am well aware,' he said in confidence to Julius Schott as he left for the station, 'how important it is for the nation that hundreds of factories like yours should be working day and night and trading with the whole world. Obviously, industry cannot grind to a halt just because of some acid smoke spilling out of a few chimneys. However, the effect on our children is more severe than on adults. If I were you, I would look into the possibility of sending your two sons overseas to some other climate, perhaps to a country like Switzerland, with mountains and healthy open spaces. For example, they could finish their school years there and it would certainly do them a lot of good. It is quite fashionable at the moment. I know many families who send their children away for similar reasons in the year before they go to university.'

Julius Schott understood the message and spoke to his wife about it, but such a radical suggestion as sending her children away was not at all to Mrs. Schott's taste. In fact, she found the proposition rather shocking on behalf of a medical practitioner.

'How can he even suggest anything of the kind?' She asked. 'As a doctor, he should come up with remedies and prescribe medicine, not try to change our way of life. You went to him for a diagnosis, not for ridiculous advice on our own children's education.'

It was quite clear that Mrs. Schott would have none of it. Nonetheless, Julius Schott did look into the cost of schooling in Switzerland. Unfortunately, as the doctor had said, it was indeed very fashionable and the prices reflected that. Many rich British children from high class families were already going there and the fees were outlandish. The Schotts were fairly prosperous people, but not rich enough to follow up the doctor's suggestion.

So, there was no real solution to the family problem. To confirm the diagnosis of the London specialist, Julius Schott also sought second opinions from other practitioners, but they were either incompetent or else confirmed the advice already received from Professor Brown. There was no doubt about it. Pollution was the root cause.

The London doctor had told the truth. And the truth, when Julius Schott started to gather more information, was pretty alarming indeed. Pollution in Bradford reached well above the levels in the rest of the country. He began to read some of the earliest demographic studies, which were just beginning to come out at that time. Adding in effects of cholera and typhus, the life expectancy of the citizens of Bradford, not just his own factory workers, was known to be remarkably short. He never revealed the true figures to Mrs. Schott, because of her sensitive nature, but he did his best, as a result of Professor Brown's visit, to organize countryside holidays far from town for both of his children as often as possible. He had got the message. It was vital for the health and even for the survival of his two sons.

He was rather discrete about it. He did not want to cause any local panic. Julius Schott was now a prominent factory owner in the town, and factories were more than ever the mainstay of the local economy. It would have been rather difficult to admit that such an important and lucrative activity presented a serious health risk for all the inhabitants.

Mrs. Schott became accustomed to the idea of travelling around the countryside. Rather than hear her beloved children coughing and spluttering all day, she accepted that it was a sensible idea to take them away elsewhere for a few weeks at a time. Doing so locally would not have been much use. The whole of the county was affected by smoke and smog. So, the holidays ranged somewhat wider, the main purpose being to get away from Bradford. The London doctor who had suggested Switzerland was right, but travelling to the Continent was too expensive and complicated for them.

Bradford had become a modern version of hell. George Augustus and George Gustavus Schott were unlucky to have grown up in such an environment. Fortunately for them, they came from a wealthy enough family to escape from it occasionally. This already set them apart from the rest of the population.

The Schott household comprised a few servants. They kept a horse and carriage of their own and lived comfortably. However, the pollution in town was the same for everybody. Julius Schott fell into the habit of reading medical magazines. He decided that Scotland was perhaps the best place in the British Isles for clean air and sent his family up North on several occasions for protracted stays. He also became interested in dieting and in healthy eating, which were not much talked about at that time. In fact, the Schott family appeared a little obsessed by medical matters according to the other factory owners in town. Julius Schott was a very modern engineer with progressive views, well in advance of his time. He cared deeply about his own workers and was regarded as a very responsible employer. Factory owners provided employment for the whole town and were looked up to by the population. In this respect, the Schotts were amongst the most prominent citizens.

Every year, the family would invite a young cousin from Frankfurt, Charles Jakob Schott, from a somewhat less prosperous branch of the family. The three young boys would play together all day and, when Charles Jakob was around, all the children spoke German together, which was why Mrs. Schott was always very keen to invite him.

Despite all the precautions taken and a rather sheltered childhood, Adolphus and Gustavus did not quite escape the bad consequences of pollution in Bradford and remained rather frail. In fact, they were to complain of chest problems all their lives. Luckily for them, both were extremely gifted intellectually.

They chose to study what was still referred to then as natural philosophy (namely, mathematical physics) for which they both had a remarkable talent. When they finally left Bradford for Trinity College Cambridge, they experienced for the first time a whole year of pure, clean fresh air and genuine English countryside. The experience was so vivifying that they promised each other never to return to the industrial hell of West Yorkshire. Their families had succeeded there, but that was not a strong enough argument in their eyes to put up with permanent pollution.

Owning a factory might be nice, but owning a factory far away and not having to go there was a lot better.

George Schott's Motivation and Philosophy

Although he would not have admitted it in so many words, the lifelong ambition of George Adolphus Schott was to be the perfect English gentleman despite his foreign origins. The quest for distinguished behaviour must have led him to consider experimental science as a vulgar pursuit, too close to machines and to the factory floor his family knew so well. Experiments seemed distasteful to him. He preferred what he regarded as important ideas, those based on a high philosophical conception of the universe. True, in British society, metaphysics was never regarded with much favour, perhaps owing to the Protestant tradition, since metaphysics seems rather close to some strains of Catholic theology. Those of the Reformed churches are less tempted by idle speculation and more respectful of the solid values of hard work. Nonetheless, in the upper classes of the nineteenth century, a slightly abstract form of thought had developed, close to the 'aesthetic' movement championed by such famous authors as Oscar Wilde.

The theory of electromagnetism, elegant and refined in its mathematical structure, with subtle discoveries such as the displacement current, had just the right mix of philosophical abstraction and universal generality to fit all the prejudices of George Adolphus Schott. It is no surprise that he chose it above all other branches of physics as his favourite pursuit. He loved its magnificent symmetry and its generality. It allowed consideration of the whole of space, abstract definitions and perfectly compact equations. What pleased him most was its mathematical rigour. Whenever he opened Maxwell's *Treatise on Electricity and Magnetism*, he felt a quiver of admiration for the beautiful capital letters representing the electric and magnetic fields, for which he experienced an aesthetic thrill known only to mathematical minds.

The beauty of Maxwell's theory was the crowning achievement of nineteenth-century physics, and it led George Adolphus to the study of quite new articles of great philosophical depth and extraordinary physical insight, published by a certain Albert Einstein. Schott chanced upon these articles almost immediately after their publication. Some of their more profound content was beyond his comprehension, but the abstract mathematics used by the author was perfectly transparent to him and he adopted it at once as an obvious extension of the principles embodied in Maxwell's theory. Gradually, he absorbed Einstein's ideas and soon placed his work next to that of Maxwell as the most sublime form of scientific thought attainable by man. The deformation of space, for example, and the role of light in the new theory filled him with admiration. He often stated that Newton and Maxwell had fought the most noble of mathematical jousts and that, against all expectations, Maxwell had won the contest. It had become necessary to correct Newton's theory, and Albert Einstein was the only man in the world capable of accomplishing such a prodigious feat. For him, this was the highest peak in the whole of physics, and what secretly gave Schott much pleasure was being able to read many of the original papers in the German language as soon as they appeared in print. This was an advantage he held over his purely British colleagues which he enjoyed to the full. He could appreciate all the subtleties of Einstein's works written in his mother tongue, measure their greatness and come even closer to the original inspiration.

In but a few years, George Schott had become one of the best theoretical physicists in Cambridge and was recognized as the young master of electromagnetic theory. He practised it as none other of his generation and had absorbed all its consequences so completely that, in his eyes, it seemed essential to the coherence of the universe. There was no question that, for example, charged particles might do otherwise than obey the laws of Maxwell at any level of understanding or in systems of any size. So, for him, accelerated charges must necessarily emit radiation and this was such a clear law of nature that it could not be violated even for the briefest instant in time. Indeed, the mere idea of such a violation seemed quite inconceivable or, as he would have termed it 'unphysical'.

During the long railway journey back to Aberystwyth after his first encounter with Rutherford and their initial discussion of the atomic model, he did not even notice, as he usually did, trees flying past the windows of the carriage. He did not smell the horrible smoke in the tunnels, did not hear the noise of the locomotive or even follow what other passengers around him were saying. His thoughts were all elsewhere. The return trip was a haze of light and shade with an unintelligible background of sounds. He was still in deep discussion at the University of Manchester and had suddenly realized that his meeting with the New Zealand physicist was a call of destiny. He had often wondered what all his studies of physics and mathematics were leading to and what purpose they would serve. Now, he knew. He had a useful and important part to play. He would explain to the world just why the Planetary Atom was quite impossible. He now saw very clearly why the whole idea was absurd.

Too many people had already fallen into this trap and even Rutherford had joined the victims. The Planetary Atom was a peculiar dream, born of the history of astronomy and had no place at all in the theory of atoms. It was a naïve extrapolation which had occurred one day to the French physicist Jean Perrin. He had just not understood the consequences of setting up a system of charged particles moving in stable circular orbits around a nucleus like the planets of Copernicus. They could never avoid radiating away all their energy by emitting light. Otherwise, they would not even obey basic rules of symmetry the German physicist Paul Ehrenfest had formulated. It was really an unfortunate illusion, a blind alley in theoretical physics. To him, George Schott, had fallen the task of proving that the Planetary Atom was impossible. It was not quite as good as discovering something completely new by himself but, in many respects, would be just as important for the advancement of science.

It is completely obvious, he thought to himself, *that the motion of a particle (the electron or any other) on a circular orbit is an acceleration. There is nothing complicated about that. Any schoolboy who has encountered Newton's theory of gravitation is perfectly aware of the fact. It is a basic rule of mechanics. So, it follows, according to a theorem of Maxwell that a charged particle on such an orbit will radiate its energy. This is written into the laws of electromagnetism and follows from them as clearly as night follows day.*

No, he thought himself as the train entered another tunnel, *there is no way of escaping this reality.*

He could see before him a long sequence of specialist papers in mathematical physics which would result from his thoughts. The instability of the Planetary Atom was the inevitable consequence of the radiation emitted by circulating electrons. This was a new form of radiation, stimulated by Rutherford's speculations. Schott would put everything back into its proper place. Scientific order would be restored.

There was no doubt: if anybody could achieve the task of restoring order and explaining the properties of a radiation which put paid to the planetary model of the atom, it was George Adolphus Schott. The problem was cut out for him. He would author a series of communications on the radiation emitted by accelerated electrons and make a name for himself. This was his physics and he was the most competent scientist of his generation to tackle the issue. He had in hand all the theoretical tools and could see exactly how to pursue every detail of his argument. In fact, this was probably the most important problem he would ever encounter. He was as happy as an artist who suddenly receives a brilliant inspiration from on high. This was his break. He would be the one to set the Planetary Atom to rest forever as a piece of nonsense.

He would of course be obliged to repeat his trip to Manchester in order to master all the arguments which might emerge from the other side. It was his duty as a scientist to listen to them all and to refute them one by one, explaining away all the errors of logic and all the false deductions which might have led great minds astray. It was important not to leave a stone unturned, since a refutation must, if anything, be even more rigorous than a discovery. He would make sure nobody, in future, would even attempt to follow this erroneous path.

There had already been a few examples in science of talented people coming up with wrong theories which had to be disproved. He remembered the mistaken ideas with pretentious names he had learned about at school. There was the mysterious flame-substance *phlogiston* from the Greek word φλογιστόν (why on earth did it have to be said in Greek?) and there was something called spontaneous generation or *abiogenesis*, again in Greek: αβιογένεσις. What a silly fashion to give Greek names to wrong ideas! At least Rutherford had not attempted that. Still, disproving yet another piece of bad science would be a kind of revenge on the old classics teacher who had bored him with endless Latin and Greek classes as a boy.

It was a bit sad, of course, for Rutherford. This, however, was not a matter of sentiment. Wrong notions had to be extirpated from science like bad weeds from a garden. Of course, in a way, he would seem a bit brutal. He would have to upset some people on the way. That was inevitable.

As this thought occurred to him, he suddenly saw the young and trusting face of Eileen before him. Yes, she would also be saddened on discovering the part he was going to play in contradicting her father's favourite idea. It could not be hidden from her. She would learn it, inevitably, from other people. No, he would not reveal it to her directly. That would be too cruel. Would she regard his role as a form of betrayal on his part? Would she ever be able to forgive him?

The pursuit of Truth, he decided, *that is the only important mission of a scientist. That is why we exist in the first place. Everything else is secondary. Once on this road, there is no choice. We must serve Truth alone and personal feelings don't come into it. Even our dearest friends and most tender thoughts cannot stand in the way. The pursuit of Truth is our religion and there is no other path.*

At a little over forty, George Schott was still unmarried, having devoted himself to a life of study. He was not much attracted to women and found no hardship in following the strict rules of Victorian society concerning morality. Oddly enough, he was the one to exert a strong attraction on the opposite sex because of his natural politeness and reserve. However, although he was vaguely aware of the fact, he never took advantage of it and would never have set out to seduce a lady deliberately. Good manners and a sense of restraint were an essential element of his education and had become second nature to him.

Of course, he had well understood (it had been made explicit anyway) the attraction he exerted on Eileen. Had he been ill-intentioned, he could well have taken matters further and, indeed, she had also captivated his imagination. But he was left with a sense of embarrassment which persisted in the back of his mind. The age difference between them was the real problem. It made a normal relationship impossible and he felt it almost as a kind of curse.

On leaving the Rutherford household after his first visits, he took away the feeling that he had found his scientific direction, but had also met the one person who, under more favourable circumstances, would have been able to bring him real happiness. To have touched the heart of Eileen was in one way a blessing, but at the same time a source of worry, because he knew from the start the sympathy they felt for each other could hardly grow into anything more.

There were definitely two reasons for him to return to Manchester and visit the Rutherfords. One was the atom, be it planetary or not. The other was Eileen. The thought of being able to see her regularly in the future was also an inspiration on his return journey. Doubtless, their friendship would come to an end some day. It was inevitable, because of tensions already in the air, but he would not be the one to cause the rift. No, he would make sure that the debate about the planetary nature of atoms would last as long as it possibly could. Nobody would know the real reason why he would spin it out.

Except Eileen, of course. She would be able to guess.

6

Ernest Rutherford
and Marie Curie

The long friendship between Ernest Rutherford and Marie Curie began a long time before he took up the chair of physics in Manchester, while he was still working in Montréal. The young New Zealander was following closely anything new to do with electrons. So, he was attracted towards the unknown structure inside the atom and the novel subject called radioactivity which seemed to hold the key to a more complete understanding of the mysteries surrounding the nature of atoms.

From their first publication, he noticed the work of the Curies and wrote to introduce himself and entered into correspondence with them. The one who answered his letters and seemed to enjoy writing was always Marie Curie.

In 1903, Rutherford came to Europe. As it happened, he was in Paris on the very day Marie Curie defended the results of her thesis work in public for the doctor's degree examination, according to the French custom. The friends of the Curies all went to dinner together after the event. Years later, Rutherford still remembered, at the end of the meal, how a sample of pure radium was passed around the table and how the lights of the restaurant were turned off so that all the guests could better admire it glowing in the dark. No doubt they had all been copiously irradiated on this occasion but, of course, nobody had the vaguest idea that there was any danger involved, so there was nobody to complain. After all, phosphorescence and luminescence were well-known and had never provoked any ill-effects.

To understand by what tortuous path the fourth son of a modest farming family in New Zealand had become a world-famous researcher exchanging ideas with the Curie couple, one has to start from his early years of struggle to achieve high education and place them in context. His biography is in many ways was just as remarkable as that of Marie herself. Being born in New Zealand was an unusual start. However, it was not a bad choice. Great Britain, in those days, was at the height of its power and influence around the world and the British Empire stretched over most of the planet. The English were everywhere to be seen, governed colonies on every continent and their trade had made them the richest nation of all.

Like all supremely prosperous and successful societies, they had developed a powerful ruling class with some well-established prejudices and opinions. Amongst these was the idea that experimental science was not a suitable pursuit for young members of the aristocracy. It was a bit of a vulgar occupation. However, there was nothing wrong with a young farmer's son taking an interest. It was even open to him to make a success of himself in this way if he proved able enough. There were a few examples of such people, including a few in universities. So, in principle, nothing prevented a talented young man from arriving in England from such a distant colony to pursue high level research. Of course, he was more likely to be from a family who had settled there, rather than some kind of aboriginal of doubtful descent. This would certainly be the case because of the structure of the education system. To be sure, not all colonies were equally probable. New Zealand had a good reputation, as the settlers who had gone there were regarded as honourable people. Australia, on the other hand, might have been viewed with less favour, as the original settlers included many criminals expelled from Britain as convicts. Thus, the country had acquired a bad reputation. Most English people regarded New Zealand as a better place to be born in. It had a nicer history. Indeed, some such prejudices still linger today.

Rutherford's father, despite having little education, was a very resourceful man and had a high practical ability which he used to extremely good effect on his farm. He had built a mill with his own hands. Having sown and reaped cereals, he could turn himself into a miller, mend his machines if required and become both an inventor and an engineer according to circumstances. There is no doubt that Ernest Rutherford inherited from his father the great innovative skills he displayed on many occasions in his experimental work.

His mother had been a teacher, and both his parents had a deep respect for study. They supported all their children in their education, but there were twelve of them, and this implied an enormous financial sacrifice on their part.

Rutherford, from the start, was a brilliant student at school, full of energy and curiosity. He was naturally gifted for the physical sciences and also impressed his teachers with his ability in arithmetic. He was an excellent rugby player, which helped the development of his outgoing personality and made him very popular with his fellow-students, as it was always the favourite national sport amongst New Zealanders. He became the best all-round student at Newton College and gained immediate entry at the local university: Canterbury College. He was soon to make unexpected discoveries in electromagnetism, a subject with which, later on in research, he developed a complex relationship, but which was the starting point of his extraordinary scientific career.

His family, of course, was very proud of his achievements, but soon ran out of money trying to support him. By some mixture of miracles and hard work, Ernest Rutherford managed to obtain the one and only scholarship available in New Zealand to support himself for five years at university. Remarkably, it was a scholarship in mathematics, which was not at all his favourite subject. Also, the support was rather meagre and he was obliged to give private lessons in order to keep going.

In 1894, he obtained his first degree in New Zealand with such high marks that he gained immediate admission to the Cavendish Laboratory in Cambridge to prepare a PhD under the supervision of Professor JJ Thomson, famous worldwide for his discovery of the electron.

Before leaving New Zealand in 1895, he became engaged to a young New Zealand girl, Mary Newton, who was to become his faithful companion. However, she had to wait until 1900, the year in which they finally got married. They were simply too poor to do so earlier.

For three years, Rutherford worked on so-called 'Hertzian' electromagnetic waves and published several high-level research papers on the subject which were much admired. In collaboration with Thomson, he used X-rays to ionize a gas and observed that negative and positive charges were separated, then recombined in absence of the X-rays to form a neutral gas again. He invented equipment to measure the speed of the ions and the rate of recombination. This experiment was the starting point of a subject he was to develop later, namely nuclear physics.

His progress was so remarkable that, after three years in Cambridge, he was appointed in 1898 Professor of Physics at the McGill University of Montréal. Obtaining a paid position at long last allowed him to bring Mary Newton to join him in Canada.

Rutherford had arrived in Montréal just after the discovery of radioactivity by Becquerel in Paris. This was a very surprising effect, dubbed 'uranic radiation' by its discoverer. It was a wonderful new subject for a young and energetic physicist who, having just been appointed to a chair, was expected to define a novel programme of research.

In 1899, Rutherford published a research paper on the ionization of air by this new radiation. He had placed some uranium between two electrodes and detected a current. This enabled him to measure the power of penetration of uranic radiation by inserting metal plates of varying thickness. By this method, he distinguished between two different kinds of radiation (he called them alpha and beta) which were both present.

At this point, Rutherford was still striving to follow a different route from the research group in Paris. So, he decided to study thorium rather than uranium, using the same equipment. He was amazed to discover a very strange effect. Just opening the door of the laboratory was enough to disturb his experiment, as though small movements of air could change everything.

He concluded that thorium produced some radioactive gas. It was enough to remove some air surrounding the thorium to change the result. He discovered that air, even at some distance from thorium, was also made radioactive and produced an electric current in his detector. However, it only remained so for about ten minutes, after which the current decreased exponentially.

That is how, in 1900, he discovered the half-lives of radioactive elements. In collaboration with a colleague from Montréal, Frederick Soddy, he established in 1902 that the gas contained radioactive elements which were not the same as the original elements of the system. Radioactivity, he announced, could cause the disintegration of elements.

This discovery caused a great stir, not to say a scandal, amongst chemists. They had always accepted that matter was indestructible and that the transmutation of elements was an impossibility. By his discovery, Rutherford was challenging the very basis of chemistry. His results, nonetheless, were beyond doubt. Pierre Curie, in Paris, also began by questioning Rutherford's conclusions, but soon had to accept them after repeating the same experiment himself several times over. Earlier, during his work with Marie Curie, Pierre had noticed that radioactive samples were losing weight, but he had not reached the right conclusion about the reason.

A lasting competition between Rutherford and the Curie couple began and was to remain the central feature of international scientific research during the whole of this period.

In 1900, Ernest Rutherford finally married Mary Newton. They had an only child, Eileen, who was born in 1901.

Two very important years in the history of nuclear physics were 1903 and 1904. The Royal Society of London awarded the Rumford Medal to Rutherford and he published a book entitled *Radioactivity,* in which he explained that the new phenomenon is independent of external influences such as pressure, temperature, chemical reactions, etc. There could be a seemingly miraculous transmutation of elements. Rutherford and Soddy estimated that the energy involved in nuclear disintegration could be twenty to one hundred thousand times as great as in a chemical reaction. Their discovery, applied in astrophysics, accounted for the mysterious source of energy in the sun and stars and could explain the stabilization of the temperature of the earth.

Of course, discovering a new source of energy could hardly go unnoticed in the age of the industrial revolution. Any engineer could not fail to be impressed by the economic advantages which would follow if it could be mastered. And if anyone doubted what the source of this mysterious new energy might be, an immediate reply came from Albert Einstein whose mass-energy equivalence was soon announced. The impact of Rutherford's work was felt all over the world.

Otto Hahn, who had, in his own laboratory, discovered nuclear fission, came to work with Rutherford's team at the McGill University for a few months. Rutherford extended his work on the alpha rays. He was able to show, by making use of electric and magnetic fields, that they were composed of charged particles much more massive than the electron.

As from 1903, he and his competitors, the Curie couple, were firmly on the march towards the most glorious period of their new subject: nuclear physics.

7

Hiring a Professor
of Physics

Professor Joseph Greenwood, Vice-Chancellor of the University of Manchester, was not a physicist. Indeed, he took some pride in understanding nothing about the sciences. On the other hand, he was extremely well-versed in all the episodes of the Peloponnesian Wars and could recite by memory long passages of the Greek historian Thucydides. As a young man, he had spent many long hours translating Xenophon's *Cyropaedia* and never failed to remind his students how important such work was whenever they complained about the dullness of his unintelligible lectures.

He was convinced that studying the classics was the best possible preparation for a successful life and distinguished career. He could refer to personal experience. In some of the very fashionable sets in Bloomsbury or even in the City of London, a few well-chosen words in Greek always enjoyed astonishing success. The governing classes were all proud of having studied Greats.

In fact, although he would not have mentioned it as it was a bit too personal, the exalted position he held as Vice-Chancellor was due in no small part to his worldly successes, and these in turn owed much to his prowess in the language of Homer.

He was, however, a little bit at a loss when the problem arose of selecting a good specialist to teach the subject of physics to the undergraduates. On reading the list of titles the course was supposed to cover, he found the material had become incredibly obscure, not to say opaque, since Antiquity and he could not blame young students who occasionally complained that lectures made no sense. His first temptation was to leave the whole matter of selecting a new professor to the specialists in the faculty and let them wrestle with such a thorny question, but he also had a nagging doubt about his own responsibility. On the one hand, why bother himself with a subject of so little interest or importance to anybody? On the other, should he try to discover more about it even if it meant wasting an awful lot of precious time? It had to be faced, physics would never help any former student of the University of Manchester walk into a truly important job (like Vice-Chancellor, for example); on the other hand, it had to be admitted that the subject was taught at university level and so could not go entirely neglected.

On the evening before the decision was due to be taken, the Vice-Chancellor had a bad dream. Perhaps he had eaten too much for dinner, or had taken a little too much port. He saw himself in the company of one of his former teachers, a man of great rigour, who reproached him a crime often committed even in Antiquity: dereliction of duty. His responsibility, he heard him say, was unquestionably the governance of the university. He had accepted it. And now, when the time had come to exercise his authority, was he going to sit back and just do nothing? Would he ever be respected as Vice-Chancellor by all those (and there were many) who looked up to him for guidance?

He woke up in a sweat after his bad dream and felt very uncomfortable about the coming day. What should he do? As he observed the dawn (sometimes visible, even in Manchester, through the industrial fog), he was filled with a new anxiety. It occurred to him that Aristotle had written about natural philosophy, a science which had something to do with the whole universe. In point of fact, he had been trained himself as a student to understand the complexities of this subject. Who could decide on it better than a former student of Corpus Christi College, fluent in both Latin and Greek, one so well-versed in Greats that all had admired him? In fact, he was really qualified to decide on anything. Why should modern physics be excluded? The very name of 'Greats' implied a path towards greatness, i.e. towards the highest positions in society. And not just any society either. This, after all, was Great Britain, the greatest empire in the world since the decline and fall of Rome (immortalized by Gibbon who, by the way was also the greatest and most illustrious historian of all time... and his favourite bedside reading when suffering from insomnia).

As he knotted his tie before the mirror, Greenwood suddenly changed his mind. No, he would not let the occasion pass. He would act. It became clear to him in his mirror, as he gave a final touch to his hairstyle, that it was indeed his duty to choose the new professor of physics. It was a question of honour. Even if he understood nothing of an area its practitioners had driven to abstraction in modern times, it was the privilege of his office. Not to exercise it would have been disgraceful.

As he sat to breakfast, he suddenly remembered the very useful story of Newton's apple. It would provide him with an excellent and irrefutable argument. He could use it to excellent effect against the most obdurate member of the scientific selection committee should any one of them dare to come up with objections. After all, despite many years of long and complicated study recounted at great length by all his tedious biographers, the great Isaac himself would never have achieved anything without the apple which fell on his head.

According to Greenwood (who felt inspired that morning) the apple did not symbolize pure chance. There was more to it — something to do with world order and some kind of divine intercession.

'The apple is a truly British fruit,' he said to himself as he finished his Eggs Benedict. Even the name of his favourite breakfast dish seemed predestined as he prepared for this important day. 'Yes, I must rise to the occasion and show my worth by acting as all my staff expect. I am, after all, the one and only Vice-Chancellor of this university.'

He reached for his top hat and frock coat and walked a few times around the small garden he enjoyed next to his private residence — one of the delightful privileges of being Vice-Chancellor — full of enthusiasm for the role he had just defined for himself. He hardly felt the light and persistent rain falling from the sky.

'This,' he decided, 'is the best moment to prepare my introductory speech. It should, of course, appear to be improvised, but I must have everything well worked out beforehand to deliver it with conviction.'

In addition to the apple, he would mention the fine example of Michael Faraday, President of the Royal Institution, that strange outfit in Mayfair, one of the finest areas in London. He had actually met him in a reception at the Athenaeum Club (a place where he could be sure no member of his selection committee would ever be allowed to set foot) and had exchanged a few words about the Latin meaning of the word: *science*. That would be a good story for the Committee. It would remind them of their humble station, of their comparative mediocrity and show them in just what circles he, the Vice-Chancellor, was accustomed to move.

For good measure, he would add that Mr. Faraday knew nothing at all of mathematics (a subject Vice-Chancellor Greenwood disliked particularly). That would quieten down a very pretentious young member of the committee who thought himself the bee's knees just because he was professor. Good to take him down a peg or two. If that lesson was not enough, he would remind him of Euclid of Alexandria, the most famous of all geometers, who never used equations but just drew figures in the sand with a stick.

'That is true elegance,' he would say in conclusion. 'None of all the clutter of cabalistic symbols some of our modern scientists would foist upon Nature.'

After his brief walk round the garden, the Vice-Chancellor returned very pleased with himself. He felt fully confident and ready to impress his audience. With such fine examples at his fingertips, his colleagues were bound to recognize that their Vice-Chancellor was up to the job and able to hold his rank. Now, he felt quite capable of confronting any of the university's selection committees in any branch of learning.

It is true that he remembered none of the names of any of the candidates who had applied and the thought did cross his mind for a moment as he entered the meeting room. The fact is that he had not found the time to look through all the applications. The job of Vice-Chancellor was a very busy one and one cannot always do everything. It was too late, anyway, to start with that. On the other hand, it was probably also quite unnecessary. He had a long enough experience of university matters to take decisions without filling his head with unnecessary detail. He would follow the example of the strategists of Antiquity and be quick to move, like Cyrus the Great on an uncharted battlefield, reacting instantly to the movements of an unpredictable enemy.

The scientists on the committee all rose to their feet when Professor Greenwood entered the meeting room. It was not so much out of respect for him personally, but after all, he was their Vice-Chancellor. They had, in fact, already prepared the ground and were pretty sure whom to choose because (as they had concluded) Greenwood was unlikely to show up and was sure to let them decide everything. The committee had studied all the applications very carefully and in minute detail. They had prepared this penultimate meeting and thought it should be a mere formality. Certainly, they were not expecting the Vice-Chancellor to attend, since he had missed all previous committee meetings.

They had already made their selection and prepared a shortlist of candidates for interview. The basis of their choice was quite a novel one as far as Greenwood was concerned: they wanted to take account of the role of the university in the social and economic fabric of Manchester and its surrounding area. They also included in their discussion consideration of the national role of Manchester as the capital city of the Industrial Revolution which, according to them, was spreading over the whole world.

'Pardon me, gentlemen,' exclaimed the Vice-Chancellor, somewhat irritated that the committee should be involving itself in what he considered to be his job, namely determining the wider role of the university, 'this is a selection committee, not a government. The role of the university is my job. I am the one who deals with major concerns of policy. Your responsibility is simply to select the five or six candidates you like best and recommend them to me. Mine is then to examine them all and to decide which one will be awarded the Chair. That is why I am here today.'

The scientists were a little embarrassed and confused. They had thought the decision was up to them since the Vice-Chancellor had attended none of the earlier meetings. This was an unexpected turn of events.

'Vice-Chancellor,' responded the professor of chemistry, who was, by nature, more troublesome than most, 'in order to establish our shortlist, we needed to take into account, not only the quality of the candidates, which is extremely high, but also the role of the University of Manchester. We cannot forget that the selection of a professor sends an important signal about us to the outside world.'

'Quite so,' added the professor of medicine. He was older than the other members of the committee and therefore felt his opinion should count more, although he did not think physics a very interesting subject. 'Our colleague is quite right. The choice we make will send a message. This is a point the committee cannot ignore.'

'I will have no question of messages,' said the Vice-Chancellor testily. 'They change absolutely nothing. You just choose the best physicists. That is why I appointed this committee in the first place.'

'Gentlemen,' said the professor of mathematics, who saw the opportunity to play a more important role by some improvised arbitration, 'I think we can all be agreed on a few points. The best candidate must surely be the one whose presence in Manchester University will also send the best possible message about us to the outside world. The choice is not difficult. It is a simple matter of logic and common sense.'

'I have no idea what you are talking about,' muttered the Vice-Chancellor, rather irritated by the doctrinaire talk of the professor of mathematics, with undertones he well understood, having encountered them before. 'May I remind you that I chair this selection committee by virtue of my administrative responsibilities as Vice-Chancellor of the university as well as for academic reasons. It is therefore up to me to steer this debate and I say it again: I don't like your idea of sending messages right and left to the world outside. That is not what we are about.'

The Vice-Chancellor particularly disliked the professor of mathematics, whom he rightly suspected of conducting a permanent guerilla campaign against the administration of the university.

'Vice-Chancellor,' said the professor of chemistry, 'I am sure none of us has in mind to challenge your rightful role in the procedures, which of course we all accept. Today's meeting is merely to choose the new professor of physics, but of course that cannot be done without defining the part he will play in Manchester. We must remember that, whereas London is our capital and the political, financial and institutional centre of the Empire, the true heart of our Industrial Revolution is here, in our magnificent city of Manchester. So, it is surely legitimate to have this fact in mind when making our selection.'

All the members of the Committee applauded such a well-phrased summary of the discussion and approved it by nodding their heads in unison. The Vice-Chancellor saw clearly that he could not press the matter further without tarnishing his own reputation. He was already the butt of criticism within Manchester, as rumours accused him of having obtained his position by pulling strings with the powerful friends he had in London. So he smiled an enigmatic smile and made an abrupt about-turn:

'That, of course, is exactly what I wanted to hear from all of you,' he said. 'And now that we are all agreed on this point, I think it is time to proceed. You have of course selected a shortlist of the best candidates. The committee must now interview them all and decide, under my chairmanship who will be the most appropriate choice in the interest of our institution.'

The members of the Committee then circulated among themselves the applications and correspondence relating to each of the candidates in the shortlist, who had been invited to Manchester for interview. The Vice-Chancellor was a bit lost in the mixture of applications, letters of support and lists of incomprehensible scientific papers.

'I think you should place them in some kind of order,' he observed. 'Otherwise, it is rather hard to know what we are doing. It seems to me that we don't need most of this information. All that really matters before the interview is an impression of the impact of their research.'

'There is no doubt,' said the professor of chemistry, with sudden assurance, 'that the best candidate on paper is the New Zealand researcher. Although he is very young, he already holds a chair in a well-regarded university in Canada. His career path is impressive and he was admitted to Trinity College Cambridge, where he studied for his PhD.'

'What! A New Zealander!' exclaimed the Vice-Chancellor, 'I don't like that idea very much. Surely we can find an English candidate to fill the position. Why do we have to cast the net so wide? Degrees aren't everything. Why bring in a candidate from the other end of the world?'

'Of course we do have other candidates,' said the professor of chemistry, 'but they are far from being as good as this one. I have looked at his work and I believe it has the potential to revolutionize the whole of chemistry.'

'We are trying to choose a professor of physics,' objected the Vice-Chancellor. 'We already have you for chemistry and we are very happy about that. We don't need two professors of chemistry, least of all one who is already spreading revolution in the subject.'

'But this man is very important,' said the professor of mathematics, 'even before he went to Cambridge and became a professor in Canada, he was awarded a scholarship in Mathematics, the only one in New Zealand.'

'That is neither here nor there,' said the Vice-Chancellor somewhat spitefully. 'I don't want another Mathematician among us. One is quite enough. Again, may I remind you that we are here to choose a professor of physics.'

'I can assure the committee,' responded the professor of medicine, 'that this New Zealander has nothing of a medical man about him. This being said, I do agree with my colleagues that he seems to be the best applicant in this particular lot. Perhaps some of his work might even find some kind of relevance in terms of health treatments in the future, although one can never be sure...'

'In short,' said the Vice-Chancellor, 'you are all proposing to appoint a candidate who started his career by touring the Colonies to impress the world. I hope you all realize exactly what you are saying. I can't imagine how this would be received, even in Manchester. What a peculiar decision this would be! French Universities, for example, never would recruit an African professor and here we are looking for one in New Zealand of all places!'

'Since you mention the French,' said the professor of chemistry, 'you must have noticed a letter from the illustrious Professor Becquerel in Paris, the man who discovered radioactivity. He also writes in support of the New Zealander. In fact, he describes him as a rising star of modern times and waxes quite lyrical about him.'

The Vice-Chancellor was about to say what he thought of French lyricism, but reflected in time that he did not know what radioactivity was and was wise enough not to comment any further as he was afraid of saying something stupid. This gave a welcome opening to the mathematician, who had been waiting for an opportunity to speak.

'Excuse me, gentlemen,' he said in a mellifluous tone. 'This brings us back to our original discussion. The city of Manchester, through our choice, can also project a strong message to the rest of the country. It is perfectly coherent and in line with our role to express a grand and novel ambition. We have here the opportunity to take the lead nationally through applications of science. Manchester is the centre of the most important innovations in the whole of Europe. Such a role, no other city can even lay claims to. That is why the New Zealander seems such an interesting candidate to me.'

'Gentlemen! Let us not rush into this!' exclaimed the Vice-Chancellor, somewhat shocked by such arrogance on the part of scientists. 'After all, we are only discussing a Chair in Physics. This is not going to change the face of the world. Why don't we start by interviewing the six candidates on our shortlist tomorrow, since we have called them all to Manchester and they will all be expecting us to see them. We will soon find out if what you say is true. The best way is to start with this New Zealander, since you all seem to like him tremendously, and arrange the others into some kind of order.'

The committee, of course, could only agree with this conclusion which brought the meeting to a natural end. The Vice-Chancellor departed, well pleased to have imposed respect for academic hierarchy and an orderly procedure.

The other professors, discretely, improvised another short meeting among themselves, because they sensed some kind of opposition to their choice as well as a strong intellectual prejudice against science coming from the Vice-Chancellor. This was unusual and quite unworthy of his high office. Clearly, he valued classics more highly than science. They could not let this pass. Some sort of counter-attack would prove necessary, but what could they do about it at such short notice? They decided that they should meet again to discuss this important matter, but did not feel able to take it further on that day. Still, they should clearly form an alliance against the Vice-Chancellor. This was becoming urgent.

Ernest Rutherford had arrived in Manchester on the previous evening. As he did not know the town, he decided to take a walk and form an impression for himself. He had done some reading beforehand and had already a high opinion of the city people called 'Cottonopolis', the Mecca of machinery and modern industry.

However, he was not expecting to see what he discovered just by walking around a few streets away from his hotel. Manchester was far from being the modern paradise he had imagined. He even began to harbour some doubts about the wisdom of settling in such a horrible environment. To convince himself, he reflected that Canada was really very, very far from the centre of the Empire and that he might not get a second chance very soon if he wanted to find a position in England. Of course, the ideal would have been Cambridge, but that was still unattainable. At least, in Cottonopolis, he would be coming closer to his ultimate goal. So, he reasoned, it was probably necessary to accept a compromise and settle for a while in a dusty and smoky place he would not have chosen to live in if he had had the choice.

Here and there, he noticed a few large red brick buildings still under construction, which were laudable attempts at improving the appearance of the town by introducing some architectural style. But, barely a few hundred yards away, other similar buildings of the same type were already covered in black soot. It seemed impossible to escape a heavy cloak of pollution all over the town which spoiled any attempts at improvement. All the walls and window frames were covered in black stains and it was clear that the clean walls still being erected were soon going to end up the same way. As he walked around, Rutherford realized that the new world which had arisen from the mastery of thermal energy was really built on a silent catastrophe nobody said anything much about. The incredible surge of industrial development around Manchester was bought at a heavy price. As a farmer's son, he knew the value of clean air and a healthy countryside. He could see at a glance that there was no chance at all of finding these near Manchester.

It seemed that coal was being burned everywhere. There was even some pride in the fact, and British coal seemed to produce a particularly unhealthy smell. Despite that, people used it for heating, for cooking and, it seemed, for everything. There was even an open fire in his hotel room and, when he tried to open the window, he found coal dust had settled on the sash and penetrated inside what should have been a clean frame. There was obviously no way to keep it out. After a few attempts to pull the window up, he had decided it was not much use anyway, since the air was no better outside.

He had read many flattering descriptions of the modern transformation of Manchester and the improvements industry had brought to Lancashire and Yorkshire. The authors were all convinced that this form of progress would soon extend to the whole of Europe and, perhaps, to the world.

He, himself, adhered to some of these ideas, but on seeing the result around him, he could not help having second thoughts.

It seemed that modernization could not be achieved without producing a lot of black smoke and soot. The roofs of houses, the hats of ladies, fine clothes and lace, nothing seemed to escape: everything would eventually be the same grey colour. Rutherford, who had seen volcanoes in his part of the world, was reminded of the ash of an eruption. Of course, volcanoes are dangerous, but at least the ash is clean and, once the danger has passed, the land in the countryside becomes even more fertile than it was before. New Zealand was a perfectly clean country.

Ash is useful, he thought to himself, *and actually helps improve agriculture. It is of benefit to the general population. The trouble with soot is that it even spoils the land. Rain will wash it into the soil and the crops will not survive. Even the trees seem unhappy in this city. It is hard to see what could be done to get rid of these bad consequences. Coal dust will eventually ruin people's lives.*

When one visits Manchester today, after the buildings have been cleaned up, the window frames renewed and the brickwork has recovered the beautiful appearance it never had for long, it is impossible to imagine what a sad place it was before. When it was one of the richest cities in the world, it actually looked awful, and Ernest Rutherford saw it at the height of its prosperity. In truth, the industrial power of a city has nothing much in common with its appearance.

Worse still than the buildings was the obvious poverty of the general population. The city was rich, but not its citizens. There were horribly deprived people wandering about in the streets looking for work but it was evident that work was hard to find and conditions difficult. Like many intellectuals and especially scientists and engineers, Rutherford had always been convinced that the world would become a better place thanks to machinery and modernization. But it seemed this was not quite what one could see in Manchester. He was surprised to discover the dire consequences of industrialization. He had would never have suspected so much misery at the very heart of the British Empire.

Maybe we need to think a little harder, he concluded, *before allowing innovations to escape from laboratories into the outside world. Inventions can sometimes be exploited so quickly that one loses control of the consequences.*

And he returned to his hotel in a state of some perplexity.

The next day, at his interview by the selection committee, the Vice-Chancellor greeted Rutherford very politely, but fired a remarkably perceptive question at him, despite its innocuous appearance.

'Professor Rutherford,' he said, 'I see that you are a much travelled person. So, please tell me: what do you think of the city of Manchester? Did you have time to walk around and gain some impressions of our town?'

Rutherford was never in the habit of avoiding any question. He always answered clearly and directly, in a loud voice. Nor did he ever hide anything of his thoughts. He described all the impressions he had gathered and the conclusions he had drawn. The professors of science were horrified. The expectations they had formed were of a candidate in love with industrial progress, a champion of modernity who would apply the very latest ideas in society at large, were all blown into the water. They winced and already started thinking about who might be the best candidate to replace their first choice.

But the Vice-Chancellor was absolutely delighted with the reply. He was impressed with the strong personality of the applicant. Clearly, Rutherford was a fearless man who would bring honesty and directness to university debates. These were wonderful qualities indeed. He was not at all the sort of candidate he had been expecting from the comments of his colleagues on the selection committee. Best of all, here at last was a scientist with a clear view of the limitations of technology, who was not blinded by preconceived ideas about progress, a young man with truly independent views. Definitely, Ernest Rutherford was the kind of person the Vice-Chancellor wanted to see on the faculty. He would keep the other scientists in check. In fact, without realizing it, he had already given the other members of the committee an excellent lesson. He would surely put down their arrogance.

The night before, the Vice-Chancellor had concocted a second question to ask, so he moved on briskly:

'Thank you indeed for such a frank reply, professor,' he answered. 'Perhaps now, I could ask you a more personal question about the academic path you have followed up to the present. I notice that you have the reputation of being a bit of a chemist, which I suppose makes sense, because there doesn't seem (to me at least) to be so much difference between physics and chemistry...'

'You are absolutely right in that,' confirmed Rutherford. 'It is also the conclusion I have reached.'

The professor of chemistry seemed somewhat uncomfortable at such direct statements.

'I see you have worked in Montréal with Professor Soddy, a chemist, who is full of praise for your skill in his own subject. But I notice that you also supported yourself for your own studies in New Zealand by a scholarship in mathematics. That suggests a lot of chopping around between subjects. Are you also a mathematician? Are you an experimenter or a theorist? A physicist? A chemist?'

Rutherford gave a loud laugh.

'I am absolutely not a mathematician,' he replied. 'That scholarship was pretty hard for me to obtain, but I had no choice. It was the only one in the country and I couldn't have continued my studies without some kind of financial support. My parents were running out of money to keep me going and they had already done a lot. That was my only motivation, believe me. Actually, I hate pure theory. If research were just about writing equations, I would give it up right away. Nature doesn't need high mathematics to follow the right path. It always chooses the simplest way. That is why simple ideas always turn out in the end to be the best ones.'

Again, the Vice-Chancellor was absolutely delighted with the reply and observed with considerable glee the effect it produced on the professor of mathematics. Out of the corner of his eye, he had the clear impression that his *bête noire* had blanched.

'Thank you indeed, Professor Rutherford,' he said, 'for such a wonderfully explicit reply. May I merely remark that I am deeply impressed with what you say and that I share your opinion entirely.'

The interview dragged on quite a bit longer because all the other members of the committee had their own questions to ask, but the Vice-Chancellor was completely satisfied. He hardly even listened to what followed. In fact, he didn't even take much interest in any of the candidates who followed.

He had decided, despite Ernest Rutherford being born in such an unusual location, that he was most definitely the right man for the job. He would be the next professor of physics at the University of Manchester. The nomination was announced on that very evening. There was no reason for any delay. Rutherford telegraphed the news to Canada the next morning.

His main problem, he feared, would be convincing his wife to move to Manchester. Being naturally cautious and respectful of her opinion, he decided to inform her in advance of all the problems he could see with the new location. It was better that she should not discover them by herself on arrival. So, he decided to spend a few more days in the town so as to find a temporary place to stay and learn a bit more about the surrounding areas.

Those few days in Manchester were not wasted and confirmed his general opinion. British society was obviously in a state of tension, with huge differences of wealth which were hard for him, as an outsider, to understand. No doubt they explained why his ancestors had been obliged to leave the country. He met a few local residents who explained their circumstances to him. Mostly, they were dire.

He had thought that modern England would have adjusted to the situation technical progress had created. He found that, in fact, there was very little social change. The rich upper classes were still in command, running everything. They had combined new machines with well-established exploitation of cheap labour and, if anything, the social divide had been worsened by the new technology. As economic progress proceeds by cycles, there were moments of full employment and others when workers suddenly lost their livelihood from one day to the next, with nothing to cushion the effects of unemployment. He actually witnessed groups of workers being shut out of their factories and heard them complain that promises of better days had been made to them which were not kept. He had a chance to study the area around the university and to choose a district as remote as possible from all the social turmoil.

On his return to Canada, he was as well prepared as possible to convince Mary and to explain what he had seen. To his great surprise, when he started to explain the problem of pollution in Manchester, he discovered that she already knew all about it.

Mary Rutherford had a taste for poetry and she had been reading the works of William Wordsworth. She knew exactly why he had fled from the town, disgusted by coal dust and the smoke generated by the industrial revolution. In return, she told him how the poet had escaped to a marvellous countryside, followed by all the bourgeoisie of Lancashire and even of London who had discovered the beauty of the Lake District through his works.

As soon as they arrived in England, Mary Rutherford decided that she wanted to see this wonderful place for herself. She almost had the impression she had been there already through the works of her favourite poet. She fell under the spell of the gentle hills, the peaceful lakes and the seemingly natural scenery. Mary Rutherford, of course, was more familiar with the really wild countryside of New Zealand and of Canada, the two other countries she had lived in. But cleverly arranged landscapes, so well tendered that the hand of man could hardly be detected, were a new experience for her and filled her with enthusiasm.

Mary Rutherford had also discovered the writings of the Lake Poets and other authors who had followed Wordsworth in his self-imposed exile from Manchester. Many of them actually belonged to the wealthy middle class who derived their income from the factories they complained about. However, it was enough to hide unsightly industries behind a few trees to satisfy their love of Nature. In fact, the Lake Poets were not as opposed to the technological revolution as they made out. They just objected to coal being burned too close to their own homes. Mary soon adopted the same point of view, one very common amongst the Manchester bourgeoisie.

She did not realize that the English Romantic school has two quite irreconcilable branches. On the one hand, the school of Coleridge and Wordsworth, ready to adapt to a discrete use of technology to make life more comfortable, on the other the uncompromising heritage of Lord Byron, the arch-enemy of all forms of scientific progress, the man who secretly inspired Mary Shelley's *Frankenstein* and who campaigned against the very concept of altering Nature. This fundamental cleavage between the 'true', but socially unacceptable, radical Romantics and the 'gentle bourgeois' Romantics, with their convenient hypocrisy, was always hidden under the carpet by Victorian society. There was no desire to sacrifice the comforts and prosperity of modern life on the altar of the absolute beauty of unpolluted Nature. So, for Mary as for so many of her contemporaries, the love of countryside never rose much above admiring the achievements of horticulture. Like many New Zealanders, she adapted to British life by becoming very conformist. Her sense of order integrated perfectly well into the rather conventional intellectual circles around the University of Manchester. Mary Rutherford was not a rebel in any way.

Had she studied all of the Lake Poets, she might eventually have chanced upon Thomas de Quincey, Wordsworth's close friend — the best-known drug addict in England at the time. This was the darker side of life in Manchester, not only amongst artists, but also in the universities and, not infrequently, ordinary people. It was the universal cure for all ills, physical or mental. Opium, and even more commonly laudanum (a toxic mixture of alcohol with opium) were on sale quite freely and often prescribed for incurable cases of consumption. All social classes were concerned and the medical profession contributed greatly to spreading drug abuse in England. Manchester was the epicentre of this unfortunate habit. It grew from the deprivation of the poorer classes, but spread widely into all reaches of society.

Mary was rather shocked to hear that one of her husband's new colleagues was taking opium. She might have accepted it from an author, perhaps, or an artist, but she was really scandalized that a professor of medicine should indulge in such a vice. To make it worse, he had been a member of the selection committee who had chosen her husband for the Chair of Physics.

'We shouldn't judge people,' said Rutherford. 'Anyway, it is understandable. This man is a doctor and prescribes drugs all day to his patients. In some cases, there is no other treatment available, and he probably wanted to try opium out on himself to understand its effects. I would say that is even the sign of a very good doctor. He just wanted to know.'

Mary was very surprised at her husband's explanation. She learned on this occasion that bottles of laudanum were freely available in certain working class areas in town. A woman was well-advised to avoid walking around in such districts. Mary immediately decided to find out where they were and to avoid them carefully.

The Rutherfords were not very rich, but did their best to live in a proper style and maintain the decorum expected of a senior professor and his family. They were often guests of rich merchants and landlords, who were proud to host a famous scientist and count him as a friend. As their circle widened, Mary discovered, in the countryside, another dark secret of English society she had never suspected. Quite by chance, as they were visiting the lands of a very rich aristocratic family, she noticed that farm labourers were housed in temporary lodgings scattered around like picturesque little villages on large agricultural domains. At first sight, they were not very conspicuous, because poverty somehow seems less real on farmland and near wide fields, under old trees or in little hamlets resembling traditional paintings of country life. But the Rutherfords could not be fooled by what was happening. They came from a farming background themselves. They knew how a farm is run and just how many labourers are needed to till the soil. So, they were very surprised to learn about farm workers and the system of tied cottages. It was strange that farm workers received better treatment in New Zealand than in the metropolis. For some reason, they would have anticipated the reverse.

Ernest Rutherford had studied in England, but Cambridge was hardly the place to discover the realities of British farming and anyway he had been much too busy with his thesis work to concern himself with what was going on around him. It is only when he returned from Canada and settled in Manchester that he began to take a closer interest in the way of life in his adoptive country.

In town, the levels of poverty were rather frightening. Even the employers who profited most from cheap labour were rather embarrassed by the sight and looked askance when they met a crowd looking for work.

There had been a riot in Manchester because of a famine, put down by a cavalry charge which had left many dead. It was known as the Peterloo massacre and was still talked about in hushed tones at that time. Today, there is a simple plaque recalling the event. This was industrial England in the times of George Orwell and Karl Marx.

In the midst of all this turmoil, engineers were fighting to change the world, and sometimes successfully so. The pragmatism of the North contrasted with the intellectual pretentiousness of Londoners. The colleagues of Rutherford loved to tell the tale of that obscure mining engineer by the name of Stephenson who had proposed to build a railway and transport passengers. The government of the time did not understand the project and sent it for comment to the respected Fellows of the Royal Society in London. This august body immediately rejected such a vulgar and ill-conceived suggestion. Quite obviously, the metal wheels of the locomotive would just spin around on the metal rails and the train would never even move forward. What the distinguished scientists in London apparently did not know was that similar engines had been pulling heavy loads of coal out of the mines for years before on metal tracks. This was a very popular story up North, where everybody enjoyed a good laugh at the scientists in London and Her Majesty's incompetent government. It was a matter of great pride for miners that the whole country was now becoming covered with railway track thanks to the most famous engineer in the land.

The Rutherfords, who in some respects were still outsiders because they came from elsewhere, were also surprised to find that ordinary people and rich industrialists and merchants lived side by side but rarely mixed. As for the highest spheres of society, the great landowners and titled nobility, they were sometimes glimpsed, but never seen otherwise than from afar. They remained a class apart, in their castles surrounded by vast domains, as is indeed still the case in a country which has long made a study of how to avoid real social revolutions. In New Zealand, the social pyramid was far less daunting.

Such was the context into which Rutherford arrived from Canada with, in his luggage, the first results of his experiments on thorium. He was inaugurating a new scene of emerging technologies based on new forms of science. He was about to change the world completely, but was rather conservative in outlook and would have preferred to keep society just as it was.

In fact, the very nature of science was about to change. It would become at the same time practical, empirical and fundamental, and would achieve all three at once thanks to him. There would be a place for theory, but a firm commitment to achieving real results. Fundamental advances would override mere speculation. The new approach was known by a special name in some countries, where it was sometimes regarded with slight suspicion. It was called Anglo-Saxon empiricism. If any single scientist was responsible for introducing this new style in scientific research, it was undoubtedly Ernest Rutherford.

The Vice-Chancellor of the University of Manchester had indeed, without realizing it, changed the course of science by appointing the brilliant young professor from Montréal to the Chair in Manchester. It was in fact a very logical choice. The consequences of Becquerel's discovery of radioactivity and the research career of the new professor all tied in perfectly with the powerful realism of the city of Manchester. It was, in fact, the perfect marriage. It is a rare event, in science, to achieve such harmony. When it occurs, it is often the sign of important things to come.

Rutherford introduced a new rigour in experimental science which had a durable influence on research worldwide. In a short time, he set up a group to rival Becquerel and the Curies in Paris. He had arrived like an explorer in unknown territory, guided by his marvellous intuition. Sometimes, he would hesitate between two possible paths. For a moment, he would seek the guidance of theorists, but then would usually take his own decision, because his first instinct was always to trust Nature above all else.

He understood and respected young talents like Niels Bohr, whose originality of thought was quite clear to him, but he was also very cagey about accepting radical new ideas. His approach, ultimately, was a mixture of intuition and common sense. He hated vague and general ideas such as the cosmologists came up with and thought them particularly stupid. On several occasions, he exclaimed: 'Let me catch nobody talking about the universe in my laboratory!'

In his opinion, theory had to stay in its proper place, which was narrowly defined, and never attempt to step out of it. It should guide but not dominate research, and experiment was to be the real source of knowledge. Idle speculation was to be excluded from science. This somewhat limited his own horizon.

As a result, the new mechanics, of which he had in fact discovered all the basic principles, was formulated, not by him, but by others who allowed theory a wider remit.

8

Rutherford's Dilemma and His Challenge to Schott

When he first met Schott in Manchester, Rutherford was hesitant. He was on the edge of a spectacular advance, but could not make it completely coherent. The Planetary Atom was a tantalizing idea, but somehow did not come together exactly as he would have liked.

The basic problem, as he was well aware, was its stability. For his Planetary Atom to exist, the electron had to orbit around the nucleus without losing any kinetic energy. Otherwise, the negatively charged electron would be attracted by the positively charged nucleus at the centre, would necessarily fall onto it and get stuck there, so there would be nothing planetary left at all. How could such a catastrophe be avoided? Why do atoms not collapse?

Rutherford had worked on electromagnetic waves in New Zealand and had perfected his knowledge of these waves in his early years in Cambridge. He was something of an expert on the subject. He knew that there was a connection between the movement of charges and the emission of electromagnetic waves. Like all matters related to charges and waves, it was wrapped up in Maxwell's theory.

Although he did not presume to master all the details of the profound mathematical work of James Clerk Maxwell and could not see quite how to sort out the complex details of a full calculation for circular orbits, his intuition told him that the answer was bound to come from principles hidden away somewhere within Maxwell's equations. It was rather frustrating that he could not find them.

Maxwell's theory is amazingly beautiful, he thought to himself, *but very difficult to play around with. Somehow, as soon as one touches it, everything seems to fall apart.*

The truth was that Rutherford did not possess enough mathematical talent to apply the full power of the theory to such a complex problem and was well aware of that. So, in fact, George Schott had arrived exactly at the right moment. Here was a great mathematician, with all the required capabilities and, as luck would have it, he was also the best expert in the country on electromagnetic radiation. This was really a godsend. Who could have asked for more? Surely, Schott would help him to resolve the matter. That is why Rutherford welcomed him so warmly.

All I need do now, he thought, *is to convince him to work on my problem. He can join my research group and we will supply him with any information he may need on the way. His expertise will be a formidable asset.*

There was already some debate underway about Rutherford's model. There were a few physicists still attached to Thomson's 'currant bun' and there were those with similar worries to those of Rutherford himself. He was sure he had done two-thirds of the work needed to establish his model of the atom, but he really needed a theorist like Schott, an authority on Maxwell's theory, for the final step. And of course, as always, there was a question of mathematical elegance. Without a beautiful formalism, it is always hard for a new physical theory to establish its preeminence.

There is no doubt, thought Rutherford, *Schott is the man for the job.*

So, one day, when he chanced on George Schott in a corner of the university library, he seized the opportunity and put the whole problem to him. Like all good problems, it carried within it a great opportunity. Schott would be able to write a formidable scientific paper describing in detail all the properties of the electron on a circular orbit around a fixed positive charge. These were properties Maxwell had not even imagined. It was a problem the great man had not worked on, one George Schott was eminently qualified to sort out. It would become one of the great contributions to fundamental physics. It would surely explain the stability of the atom.

Schott listened very carefully to what Rutherford had to say. He had followed much of this path already, though differently, but he kept many of his objections to himself, out of his natural British politeness and reserve:

'I was aware,' he said eventually, 'that you might have a project of this kind. In a way, you had already broached the subject in one of our earlier conversations. Also, I have read your papers very carefully and I know what conclusions you draw from your observations. However, I fear there will be some quite fundamental objections to your proposal and they seem very difficult to overcome. An electron circulating around a point centre is subject to acceleration. That is just elementary Newtonian dynamics. And accelerated charges radiate. That is also a fundamental principle in Maxwell's theory. So, there has to be a loss of energy. I really can't see how to avoid this consequence. To my mind, it is an inescapable conclusion.'

'That is what you think now,' objected Rutherford, 'but you have not actually worked the full problem out. Nobody has. And I don't believe in bandying about general theorems when the particular case has not been solved. Especially when there is clear experimental evidence which is contrary to what you say. So, this theoretical problem needs to be addressed. Why do you think your objections are so fundamental? This is just your impression before having done any serious work on the matter.'

'With respect, it is more than that,' answered Schott. 'The question of radiation by an accelerated charge is a fundamental theorem of our great Maxwell. Any charged particle must obey this rule. I can see no way out. This means that the model you are proposing cannot be stable. If it were, this would contradict the very basis of electromagnetism.'

'I don't see that,' said Rutherford, irritated by Schott's opposition. 'I have observed the structure of the atom, so I know it exists. Just appealing to fundamental theorems without relating them to the problem at hand takes us nowhere. What you need is to go for the detail. It is very likely that there is something hidden in there which you have overlooked. You theorists contemplate Nature from a distance. I get as close as possible to it and watch its forces at play every day. So I can tell you exactly what is happening. You need to keep an open mind.'

'Do you have doubts about Maxwell's theory?' said Schott with astonishment.

'Of course not,' answered Rutherford. 'Nobody has any doubts about Maxwell. That isn't the point. I think what is wrong is the conclusion you are drawing from his theory. That is all. There must be a flaw somewhere in your argument. Forgive my saying so, but you must have forgotten something. Maybe it is a question of symmetry. The atom probably can't radiate because of its shape. You must take its configuration into account. You need to consider its charge distribution. There must be a geometrical effect, a circulating assembly of charges which cannot radiate. Try looking at the problem from a different angle. Go through it all in detail. Look at the Poynting vector, which is supposed to describe all the radiated energy. There are so many tools you theorists have available... Why don't you apply them?'

As a former Cambridge scholar, Rutherford of course was in awe of the great Maxwell, one of the gods of nineteenth-century physics, and would never have dared to challenge arguments clearly based on proper implementation of his theory. He knew that George Schott was regarded by all the experts as the most highly qualified practitioners of Maxwell's theory. So, he was the best person to find a way around the problem.

'Very well,' said Rutherford, 'we are not in agreement on one point, but that is an open question and the debate will be useful. So, I still think it is appropriate for me to ask you to solve the real problem, with its circular orbits, exactly as in the Planetary Atom, and come back to me with the requirements theory imposes for them to exist. Just consider the system I have proposed and tell me what is needed to make it work.'

'You can use all the equations you like, from Maxwell or elsewhere. Just throw the book at it. If there is radiation, as you say, then I want to know what kind of radiation it is and what its properties are, so that we can look for it experimentally. I don't want any general arguments. I want the specific case of circulating electrons. Take the very simplest orbits to make your calculation easier. I stick to my position that the system can't radiate. If it does, according to your theory, then the theory must be wrong, because atoms are obviously stable.'

'Very good,' responded Schott, 'I accept the challenge. I will write you a memorandum on this particular problem and give a full account of how the acceleration of electrons on circular orbits induces radiation. I will consider the properties of this radiation and how it might be observed. In fact, I will translate your Planetary Atom into rigorous equations so that you can see for yourself what its consequences are in line with Maxwell's electromagnetic theory.'

'That is what we shall see,' retorted Rutherford. 'At any rate, it is the best approach to sort out our disagreement. Just calculate what I have proposed and nothing else. After all, Newton solved a very similar problem for one planet and one sun. This is really the simplest configuration. I would bet there exists an exact solution, maybe even a very simple formula, something completely straightforward you will easily find. Afterwards, it will seem completely obvious to all of us and we will all wonder why nobody thought of it before.'

'I quite agree,' said Schott. 'The problem you have described must have a simple solution. It will be rather more complex than Newton's problem because gravity does not radiate. In your problem, there is the magnetic field to consider. However, we have a perfectly well established general theory, so there has to be a complete solution.'

'It's a deal!' said Rutherford. 'Do the theoretical bit and get back to me when you have worked out a proof. Let's wager a bottle of whisky either way to the one of us who turns out to be right!'

George Schott was not a whisky drinker, but immediately agreed. In fact, both of them were very happy with the outcome of the conversation — Schott, because he had managed to convince Rutherford to listen to his objections and Rutherford, because he had convinced Schott to work on his problem without conceding any ground concerning the validity of his atomic model. Recruiting a theorist as important as Schott was not an easy task and this in itself was a success. The outcome, he felt sure, would be some new discovery, a subtle theoretical twist which would allow his Planetary Atom to exist without any radiative losses. He was pretty sure Schott would find the answer. Indeed, if Schott had been the only opponent to worry about, the situation would have been simpler. Rutherford, however, was having to battle on several fronts.

Somehow, the scientific community had smelled a rat in his model. His rivals were out to rubbish his work, and this seemed to be the chance for them to do it. Contradicting Maxwell was a pretty dangerous line of argument and Rutherford was beginning to feel uncomfortable about his own position.

Rutherford and Marie Curie often exchanged private correspondence about their work. Rutherford's opinion about female scientists was a bit ambivalent, but he knew he was in a dialogue with the rest of the group in Paris, which made it worthwhile. He also had to be a little careful how much he revealed about his work since they were his prime competitors. Marie Curie was always very forthright in expressing her opinions and this at least made it safe. The opinions from Paris were useful in many ways and sometimes the criticism was valuable too. Pierre Curie was a fine physicist, and the New Zealander had a lot of respect for his ability. As it happened, the Curies did not seem to have any ambition to work on atomic structure, so this was a subject Rutherford eventually decided to tell them about, as he was interested to hear what they would think of his Planetary Atom. As he anticipated, the response from Paris was pretty negative and rather strongly worded:

It is possible, wrote Marie Curie, *that your model has a grain of truth in it, but I confess I am not convinced. You keep on quoting the 'currant bun' or 'plum pudding' model of your former supervisor Professor Thomson as the model which would be most naturally opposed to your own. This is because you see everything through the eyes of the Cambridge school you come from. In reality, there are many other precedents concerning atomic structure. For example, you forget the French physicist Jean Perrin, who proposed a long time ago an atom like a solar system in which the planets would be the electrons. So, your model is not at all the first one. And there are plenty more. The German competitor of your JJ Thomson, Philipp von Lenard, for example, has proposed his theory of 'dynamids' according to which electrons in the atom do not move freely, but are each attached to an individual positive charge. I know that you have an intense dislike of this horrible man (as I do also) and that you term him the monster of Heidelberg, despite his Nobel prize… but sometimes even such unpleasant people can come forward with good ideas. He may be odious, but he is no fool when it comes to physics. Then you have the Japanese researcher Hantaro Nagaoka. He has a very picturesque model which is also a bit like yours, inspired by the planet Saturn. He also says that the positive charges are concentrated in the centre, but has the electrons distributed in a plane like the rings around this planet. Then, you have Johannes Stark with his 'archions' which are rather like little magnets. They are rather cleverly arranged in circles so that they can rotate without ever radiating. That seems to address something you have neglected in your model. And last of all, but by no means least, there is the extended electron model of your compatriot George Schott. You make no mention of him, which surprises me because, like you, he is from Cambridge. Did you never meet him?...*

Although he accepted criticism of his ideas, Rutherford was rather irritated by the tone of Marie Curie's letter. She seemed to be suggesting that he had not done all his homework properly and that he had left out the contributions of many other scientists from the bibliography of his paper. So, he penned a slightly acid reply, in which he attempted to clarify, in particular, his relations with George Schott. True, they had both studied in Cambridge, at Trinity College. However, they had not been contemporaries. Since then, they had indeed met, and regularly kept in touch, but Schott was teaching in Aberystwyth, a provincial town in the depths of Wales, several hours by train from Manchester and from Yorkshire, sometimes involving a full day of travel to get there. After a moment's reflection, he added another sentence to explain that he, Rutherford, was in favour of JJ Thomson's description of the electron as a point charge and did not believe in any theory of an 'extended electron'. Thomson was surely right to consider the electron as a tiny and very light particle. Many experiments by Thomson as well as by Philipp von Lenard established beyond doubt that the electron could easily pass through the narrowest of apertures. An 'extended electron' could never do that.

Rutherford thought that would be the end of it, but Marie Curie was very persistent and soon responded.

You must admit that there are many more existing models than Thomson's 'currant bun' as alternatives to yours. Your experiment on the scattering of alpha particles does, I agree, say something specific about the concentration of positive charges in the centre, but this is not enough to allow you to dismiss all the other proposals. In particular, I repeat, the ideas of Jean Perrin and Hantaro Nagaoka are perfectly compatible with your observations, so your experiments do not disprove them…

Rutherford couldn't help asking himself, on receipt of this second letter, whether he had been wise to start a discussion with that Frenchwoman. Once again (it was not the first time he had noticed this trait of her character) she was proving particularly obdurate in her opinions. Her persistence in some situations could be a quality and he recognized that. It had enabled her to keep going without ever weakening in her quest for the hidden element radium, as everybody knew. Yes, she had worked herself to death under atrocious conditions. However, when it came to ideas, her obstinacy ceased to be an asset. Her strength of character could actually become a definite nuisance. Marie always seemed to choose the wrong camp in a discussion and never wanted to give up arguing.

Still, there might be some point in answering her letter, so he jotted down a few additional points on the back of the envelope. First, he noted that Marie Curie's examples were all drawn from speculative theories without any basis in fact — that is to say: in experiment. Jean-Perrin's cosmological analogy could hardly be called a model. If it were, then along similar lines, the ancient French philosopher Blaise Pascal had said pretty well the same thing centuries before in his famous comment on the mites crawling around in Dutch cheese. There was a much more pertinent (and quantitative) comment by James Jeans. He had discussed all the previous models of the atom in detail and performed a very clever analysis. He showed that none of them was of any use because they could not yield a frequency, which was very necessary in order to explain the observations of the spectroscopist Fraunhofer. In fact, Jeans had concluded that a new model was necessary to account for stable atoms of the type he, Rutherford, had found, a model which would also explain spectral lines.

The key to everything was obviously observation. Just making up more models was no better than stamp collecting.

This last remark was Rutherford's usual way of drawing any discussion to a close. Stamp collecting was, in his view, the ultimate example of futility. So, he felt he had now drawn the whole matter to a conclusion and was sure Marie would find nothing further to add on the subject.

To his annoyance, a few weeks later, he received Marie's reply. She would never give up. Obviously he had, as usual, underestimated her fighting spirit.

I wonder, she wrote, *whether you are not underestimating the power of theory. Have you read the mathematical paper by Paul Ehrenfest? He describes a very general approach. In it, he claims one can construct distributions of charges and currents consistent with Maxwell's theory of electrodynamics, which are able to rotate around a charge centre without radiating. This could explain the stability of atoms. In fact, he has even worked out the general condition rotating charge distributions must satisfy in order not to radiate. Surely, amongst these, must be one which would account for your observations. Maybe this would be a more relevant approach than your 'Planetary Atom'.*

This was so far-fetched that Rutherford decided in the end not to answer. Of course, he knew the paper Marie Curie was referring to, but it was just a piece of abstract mathematics he had judged far too abstruse to be of any real use. He had often encountered this type of paper. Authors would seize on some much too general idea and cast it into a sophisticated mathematical formalism which just clouded the issue. To be convincing, any argument had to lead to hard facts. He wanted real predictions, which could be tested. As for the paper by Ehrenfest, it failed his first criterion: could it be summarized in a few words non-specialists would actually understand? He didn't try the barmaid on Marie Curie but was half-tempted to do so.

This German theorist, he concluded to himself, *may think he is very clever but is missing the point. Nature loves simplicity and would surely not follow the tortuous path of his reasoning. He is just writing another paper to please an audience of mathematicians. I must ask Schott for an opinion, but I don't believe a word of Ehrenfest.*

By chance, Eileen was sitting next to him when he opened the last letter from Marie Curie. He felt an urge to talk about it and vent his irritation at the obstinacy of his correspondent. So, he explained in a few words the latest arguments about his Planetary Atom and told Eileen that Marie Curie was raising a lot of objections by citing Japanese, French and German studies.

'You see,' he added by way of conclusion, 'scientific research can sometimes be very annoying. We have endless wrangles between experts and each one of them always believes he and only he can be right.'

Eileen was very surprised to be involved as a witness. She discovered with some amazement that there were people in distant lands (and of course always this same woman) asking themselves questions about atoms rather similar to the ones debated in their home until late at night. This was a bit strange. Her intellectual universe contained three fixed locations, forming a kind of triangle. There was of course Manchester. Then, there was Montréal, where some of her father's former colleagues were still working. Then, there was Cambridge, naturally enough, since her father had studied there and it was the font of all knowledge. She had been obliged, somewhat reluctantly, to include Paris, because of the fact that Monsieur Becquerel had discovered radioactivity in a drawer of his desk, but this was somewhat accidental, although Madame Curie had taken advantage of that fact. Now, Germany and Japan were becoming involved. Eileen didn't understand anything about the objections the difficult lady from Poland was making, but one thing she did realize was that this Planetary Atom had made everything enormously complicated.

She also discovered something quite interesting. Her new friend George Schott was obviously an important person in his own right. He had thought up some ideas of his own which Marie Curie knew about and had referred to. It was a bit surprising that a woman should stand up to her father in a scientific discussion. This was not really very feminine and, to say the truth, on the borderline of bad behaviour. Many people seemed to think highly of her, but she was obviously rather a pretentious person and did not know her place. Surely, if she had good manners, she would have found more polite ways of expressing an opinion than writing such letters. For example, she might have mentioned that the remarks were not hers but made by someone else. That would already have been more tactful.

Eileen stopped at that, but clearly she had in mind that the 'someone else' should really have been a man. She had been well brought up by her parents. Later on, when the compromising affair between Marie Curie and Paul Langevin came out into the open, Rutherford was very careful in his reaction and especially cautious to shield Eileen from idle talk about such a sensitive subject. Great Britain, after all, was still a very prudish society.

Indeed, the Langevin affair threw Rutherford into something of a quandary. What should he do? Should he still communicate with his erstwhile competitor and occasional collaborator? Despite a natural sympathy for Marie and some admiration for her courage, he did worry about the right course of action. Should he stop writing to her? He decided to be more guarded in his correspondence from then on. To his American friend, Bertram Boltwood — who had been a colleague at the University of McGill, and who sympathized with his misogynist opinions about Marie — he revealed his innermost thoughts about her, which were rather negative. The two of them agreed that, after all, Marie Curie had not displayed so much imagination in her research. She actually owed her success, they wrote to each other, to perseverance in a rather daunting task nobody else would have wanted to take on. She had a lot of guts — which accounted for her success — but no natural talent. In short, as far as Rutherford was concerned, physics was not really a subject for women.

On another occasion, when he met Lise Meitner in Otto Hahn's institution, he merely noted: *Lise Meitner is a young lady but not beautiful, so I judge Hahn will not fall victim to the radioactive charms of the lady.* He made no mention at all of her scientific ability, which obviously made no impression on him at all.

Luckily for him, he had a good relationship with Marie Curie, for whom he had some respect although he could not help some backbiting in letters to other people. All his life, Rutherford retained a degree of Victorian misogyny which surfaced only in his correspondence with like-minded colleagues.

A case in point occurred when Marie Curie published her own treatise on radioactivity. Rutherford, of course, had written a book himself with pretty well the same title and was toying with the idea of a second edition when Marie's book reached him. He was a bit surprised to see her work and, as a result, gave up his own plan. Again, he wrote to his friend Boltwood: *Reading her book,* he confessed to his correspondent, *I almost had the feeling I was reading mine, with just a few additions to represent the work done in the last few years. Once again, the poor woman has done an enormous amount of work. She has written a treatise which will probably be useful for a couple of years, but doesn't add anything noteworthy. It will help a few specialists who want to save themselves the effort of looking up original references in the library; but her work will soon be overtaken. I wonder whether it will have been very productive, even for her.*

Rutherford's view was basically that a woman could only play the part of a housekeeper in science. He also objected to Marie Curie's obstinate quest for useful applications. For him, physics was an art form, like poetry and, as the best poets are supposed to be those whose work is totally useless, he elevated research to a plane at least as high, where applications could add nothing of value.

On the subject of radioactivity, he took the view that: *the energy produced by the atom is a very small kind of thing. Anyone who expects a source of power from the transformation of these atoms is just talking moonshine.*

Einstein had accepted Marie Curie as an equal and had been very clear, asserting publicly that humanity owed her one of the greatest discoveries as an independent scientist. Rutherford, just like Boltwood, found this hard to accept. He remained convinced that she had been guided at every turn in her discoveries by Pierre Curie. In his view, without his assistance, she would never have made such spectacular progress.

Fortunately, Albert Einstein had it right, and his view is accepted by all today. In the meantime, in spite of Ernest Rutherford's disapproval, the medical applications Marie Curie had invented took hold all over the world, giving rise to a wide network of 'Radium Institutes'. Rutherford's view that university science could sometimes introduce new ideas into society but should not become directly involved in developing applications were turning out to be wide off the mark, at least by present-day standards.

*

Chapter 9

Two Gifted Brothers
in Cambridge

George Adolphus Schott was very attached to his brother, George Gustavus. He had a dream that they would climb intellectual mountains together in physics and in mathematics. They had started brilliantly on this course, both of them gaining admission to Trinity College Cambridge, so this aim seemed more tangible than a mere dream.

However, fate decided otherwise.

As George Adolphus continued making progress in his own studies and absorbed the great works of earlier scientists, he enjoyed sharing his newly acquired knowledge with his brother. They took pleasure in discussing what he had found and in sharing the joys of discovery. This, he thought, would gain a lot of time for George Gustavus and bring them both to a high peak of understanding.

The brothers Schott were often held up in the university as a rare example of an intellectual tandem, in the style of the brothers Johann and Jacob Bernoulli, the two famous mathematicians. Great things were expected of them both, working in harmony as a family team.

However, although Adolphus was making huge strides, Gustavus seemed to be gradually losing momentum. His motivation was failing, the sharpness of his mind seemed to be decreasing and his mathematical agility was declining day by day.

Initially, Adolphus took this as a sign of tiredness and concluded that his brother had been working too hard. He just needed a bit of rest. So, he was careful not to push him too hard in their joint efforts. The pair were physically and mentally very similar to each other and he thought he understood his brother completely. However, as the days went by, he began to perceive something more serious. Gustavus seemed to be suffering from a strange mental problem he could not come to grips with.

Adolphus then tried the opposite approach. Maybe he had given his brother wrong advice and should never have suggested rest. Perhaps the right method was to keep up the pressure. Gustavus always followed, or tried to follow, the advice his brother gave him. But the result was again quite different from what Adolphus had hoped. Instead of recovering his mental faculties, he was sinking yet further. He would remain all day confined to his rooms. When Adolphus returned, he would find him sitting in front of an open book without even turning the pages over. He seemed to be lost in thoughts he could no longer express.

When this happened, Adolphus left him for a few hours and, on his return, found him still dreaming in front of the same page. Apparently, he had not made the slightest progress. Some months later, as there still seemed to be no improvement, Adolphus became really alarmed about the condition of his brother. He realized that he could no longer keep the matter secret and decided to speak about it to his tutor, a man of long experience. He described all the symptoms to him and explained why he had not felt it appropriate to draw attention to the problem until then.

'I have come across this kind of situation before,' said the tutor. 'It is a rare condition, and seems to affect mathematicians more than other people. It usually hits very young and very brilliant students, but nobody knows why. They go through a fallow period where they seem to lose interest and sometimes turn off completely. There is not much one can do. If you are lucky, he will just snap out of it at some point and recover. It is impossible to say. The only thing we can do is wait. If there is any remedy, it is time, so don't try anything which might upset him. Give him a chance and he may get better. Don't apply pressure of any kind. Definitely not.'

Adolphus was rather embarrassed to have spoken out and revealed the problem, but on the other hand it was a relief that someone more

competent and experienced was willing to take charge. In view of the
small number of students at Trinity and their very high quality, they were
treated as the national elite. Their well-being was primordial and all the
professors took the greatest care of their health, both mental and physical.
So, the case of Gustavus was a matter of great concern, and the decision
about him was not to be taken lightly. However, after a few months of
observation, it became clear that the university could do nothing to help.
The time had come to send him back to his family in West Yorkshire.
No doubt they would be able to take care of him and, perhaps, nurse
him back to a stable condition. There was very little chance at all he
would ever recover the mathematical abilities he had displayed on his
arrival in Cambridge.

Adolphus knew very well what the decision implied for his young brother.
Of course, the family would look after him. There was no question of
letting him down. They had the means to protect him from the outside
world, except in one respect. Returning to Bradford meant that he would
be back in the middle of toxic fumes, breathing in the air they had both
vowed to leave behind. For Gustavus, there would no longer be any
escape.

So as not to disturb the studies of Adolphus, Julius Schott travelled from
Bradford to Cambridge to come and fetch Gustavus and bring him back
to Yorkshire.

Adolphus found himself suddenly alone, without the partner he had
dreamt of raising with him as a collaborator into the higher spheres of
scientific learning. He had lost his intellectual twin and was obliged to
abandon him to a slow and depressing state of mental decline. It was
almost worse for Adolphus that he would not witness events in Bradford.
He had a feeling of guilt and woke up in the morning wondering about
his brother's return to the family home, surrounded by its cluster of
smoking factories still poisoning the whole neighbourhood.

From that time onwards, Adolphus had strange dreams: he sometimes saw his brother returning to Cambridge. He had undergone a miraculous cure. He would be resuming his studies of mathematical physics and they would be able to work together again as before. For years after Gustavus had left Trinity for Bradford, Adolphus had the same dream in which his brother would recover and return, full of enthusiasm, so that their happy collaboration was restored.

However, every time, he would wake up, still the solitary student, ploughing the same furrow alone, with nobody else by his side.

From time to time, there would be a letter from his family. Such letters, however, did not bring good news. Gustavus was pursuing his inexorable mental decline. Nothing seemed to interest him any more. The symptoms observed in Cambridge had worsened. He still recognized his relatives when he saw them, but seemed to have lost any ability to think for himself, spoke very little and never left the house alone. His condition had become so serious that the question of finding a place for him in a mental home was raised. However, in those days, there was a lot of talk of bad treatment in such institutions and Julius Schott could not bear to think what might happen to him there, so he decided to keep his son in the family home whatever his mental condition.

For Adolphus, the experience was traumatic. He developed a fear of mental disease which led him to become quite solitary. There were days when he harboured the fear of going the same way as his brother. He wondered whether the world around him would always be accessible, or whether he might gradually lose touch with reality. It had never occurred to him that he might have to continue alone along the path towards deepening his knowledge. This was unexpected and rather frightening. He also had the feeling he could not discuss the matter openly with other people. They would immediately suspect him of being fragile and, perhaps, on the verge of losing his mind. He knew that there were cases of genetic dementia, passed on from generation to generation in princely families. He had heard the story of a German aristocratic family, the House of Hesse-Darmstadt, and knew of similar cases in the Wittelsbach family. He began to suspect that his brother's problems might also visit him one day, should their origin be hereditary.

He had always been rather secretive about himself and very rarely revealed his innermost thoughts to anyone. The misfortunes of his brother Gustavus created a new barrier with the rest of the world and even with those closest to him. As a result, during his studies, he became something of a recluse.

Chapter

10

Why Ever Aberystwyth?

Reputedly, mathematicians have no sense of the realities of life. Nonetheless, it may seem strange that a national authority in the important subject of electromagnetism, George Schott, should have chosen a form of British intellectual exile by withdrawing to such a secluded place as Aberystwyth. It was completely off the map at the time. To many of his contemporaries, the Welsh name of this little town was so difficult to write or even to pronounce that its spelling remained a bit of a conundrum.

However, for him, there was a very good reason for choosing such an unusual location. It was not simply an amiable eccentricity on his part.

The little town of Aberystwyth, when he settled there, was one of the most isolated places in Wales. The nearest significant conurbations were Swansea, 89 km to the South, Shrewsbury, 100 km towards the East, Wrexham, 101 km to the North and the capital of Wales, Cardiff, 122 km away. More importantly for Schott, who occasionally needed to travel for his research, London is 290 km from Aberystwyth. Settling in such a remote place was therefore something of a challenge, and did occasion a lot of surprise amongst his contemporaries in Cambridge.

Of course, anyone familiar with his childhood would immediately have understood the choice. He had grown up in the dust and soot of an industrial town. His health was still precarious and left him with bad dreams of toxic fumes. When his colleagues smoked, the habit would cause him great anxiety and he usually found a pretext to leave the room. Specialist doctors he consulted from time to time were unanimous in their advice. Because of his childhood in West Yorkshire, he needed to live far away from all forms of pollution, in the countryside or by the sea.

‘Whatever you do,’ they would all say, ‘never return to live in an industrialized city and keep well back from open hearths when indoors.’

And some would add:

'Short of going abroad, your best bet would be to move to the West Coast, where the wind blowing in regularly from the sea and the ocean could restore your health.'

He had heard the comment so often that it stuck in his mind. In Victorian times, the seaside town of Aberystwyth was sometimes referred to as the Biarritz of Wales because of a certain similarity of climate. There was a degree of exaggeration in the comparison, but there was also some truth in it. Windy and stormy weather coming in from the west brought fresh air from the Atlantic and the town was hundreds of kilometres away from industrial activity. For George Schott, settling there was a hugely attractive proposition. He regarded it as life-saving.

What made it possible, however, was the new Cambrian railway line which, from 1870, connected Aberystwyth to the rest of the world via Machynlleth and Carmarthen. There had been a good deal of investment in the construction of the seafront and the little town had great ambitions of becoming a posh seaside resort. The dreams of greatness lasted long enough for beautiful villas to be built and one speculator went so far as to finance the construction of a gigantic venture, the 'Castle Hotel' — a rather crazy Victorian folly with turrets and battlements worthy of a Bavarian king. Its plans were so ambitious that the building was never completed. Its promoter was ruined by a financial bubble which burst in 1865, leaving a huge but empty architectural shell in the middle of the town.

The building stood abandoned for quite some time and was eventually bought up by a well-meaning society which had come into existence with the sole purpose of creating a local university. That is how, in 1872, the University College of Wales was founded and soon became known as the University of Aberystwyth. At its inception, there were only three professors and twenty-six students. As for the buildings, they remained unfinished for a long time and had a chequered history.

When George Schott heard about this new university, it struck him immediately as the ideal solution to all his health problems. It was exactly the institution he needed in order to pursue a research career, and so he applied as soon as he was able. For the authorities in Aberystwyth, attracting such a brilliant scientist from the University of Cambridge seemed almost miraculous. Thus, everybody was happy when he was appointed as a new lecturer, soon promoted to professor.

George Schott had found the perfect location, where he could work away peacefully in the seaside air. Very occasionally, he needed to travel to London and to Manchester, but this was now possible at amazing speed due to the novel railway network. True, he would be plunged into smoke and dust for a few days and would return to Aberystwyth with chills and coughs, sometimes even a touch of bronchitis. But a few days by the seaside would work wonders and soon effect a miraculous cure.

I will never be grateful enough to the clever doctors who advised me to settle here, he would always tell himself as the train approached Aberystwyth station. *Without them, I would never have thought of coming here, and they were absolutely right. This is undoubtedly the best place in the world for my lungs.*

He had even become attached to the peculiar semi-medieval university building — an example of what is sometimes called the 'picturesque' style. Unfortunately, the original construction had been damaged by a fire, due to the explosion of an oxygen cylinder in 1885. The damage had been so serious that the university was very nearly closed down and it took a public subscription in Wales to keep it going and partially rebuild it in 1896.

In Aberystwyth, Schott began his work on the problem of the electron orbiting around a centre of force. His plan was to write a treatise on the subject which would encompass all that could be deduced from a full implementation of Maxwell's theory, without any compromises or approximations. Schott was unhappy with the way many physicists proceeded. He did not like piecemeal solutions. This was a problem which deserved to be treated properly. As he advanced and considered it in all its detail, he became struck by the beauty of the whole construction. It also had its turrets and battlements, strangely complex and decidedly beautiful. There was no denying it: the problem of the circulating and radiating electron, which Rutherford had refused to envisage, was the most subtle dream one could conceive for the full implementation of the theory of electromagnetism.

Soon, he worked out why the radiation had to be linearly polarized in the plane of the electron's orbit and elliptically, in opposite senses, above and below, leading to the two opposite circular polarizations along the axis of rotation. An extraordinary and admirable consequence of the geometry of the system! This, in itself, was a most beautiful result.

As he proceeded, the pages he was writing filled up with new equations. Each seemed more elegant than the previous. He had to pause from time to time to provide his future readers with a few words of explanation about what he had achieved, but always did so with a feeling of regret, because the mathematical structure he was building was so much more perfect than any of the words he could find to describe it.

Occasionally, he would take a walk along the seafront to breathe in the fresh air and an ordered ring of electrons would rotate in his mind as he went. He now saw them as forming bunches, regularly spaced around a circle, equidistant from each other. They would all rotate together, at the same speed, around a common centre. He was able to work out the power of the radiation emitted by the whole assembly, its angular distribution, and the details of his calculations revealed a magnificent natural order of electric and magnetic fields. In itself, his solution was a tribute to the wonderful structure of the mathematical theory invented by the immortal genius Maxwell.

George Schott was full of enthusiasm for the beauty of this unique problem. As he developed its amazing richness, he even forgot that he was destroying step by step — or so he thought — the Planetary Atom of his friend Ernest Rutherford. This was not destruction at all. It was creation. One day, as he pressed forward with his full description of the circulating electrons, he suddenly realized that he could build into it several entirely new properties he had become aware of by reading the papers of Albert Einstein. These papers were extremely difficult to understand fully and he did not claim to have absorbed all their consequences but the equations, for Schott, were at least simple enough to apply and predicted something very novel and interesting. As the speed of the electrons approached the speed of light, they would turn so quickly around their centre that the emitted radiation would become concentrated into a narrow cone, a pencil of light which would sweep around ahead of them, on a tangent to the circle of the orbit. The extraordinary simplicity of this result was one of the finest details he had discovered to include in his treatise. Schott was understandably very pleased with himself on the day he found it. This was a lovely result and the elation he felt was so exciting it kept him from sleeping three nights in a row.

After several months of continued toil, he had accumulated an impressive amount of material, all of which was new. He now felt confident that he had treated the full problem and would be able to persuade Rutherford to give up his rough and ready speculation about the Planetary Atom. His treatise was now ready and he would submit it to him as the ultimate proof that the half-baked idea could not work.'

As promised, Schott put the manuscript of his great work on the radiation emitted by circulating electrons into a small briefcase and caught the train to Manchester. He wanted to deliver it personally to the man who had inspired it. Although he realized reading it might be a disappointment for Rutherford, Schott felt very grateful to him for having invented such a beautiful problem. So, as he left Aberystwyth, he reminded himself that he should also thank him for the wonderful opportunity his idea had provided. Finding great problems which can actually be solved is one of the most important contributions which can be made in science. Schott was well aware of that.

Upon his arrival in Manchester, Schott went straight to the university to deposit his treatise directly on Rutherford's desk:

'I am sorry it has taken me some time,' he said, 'but the problem is a beautiful one and I wanted to do justice to it.'

Rutherford was somewhat taken aback by the mountain of paper which had landed suddenly on his desk.

'Why did you have to write so much?' he exclaimed. 'Couldn't you make it a bit shorter?'

'There was an awful lot of important detail to work out,' said Schott, 'and it is such a fundamental problem of physics that it has to be tackled head on, without any approximation. The beauty of it is that there does exist an exact solution, as you had suspected.'

'I am not sure one really needs so much information,' responded Rutherford, 'I always say that a good idea is one you can explain even to a barmaid. If she doesn't get the point immediately, then the idea was hardly worth the effort.'

Schott had heard that comment before from Rutherford and was well prepared for it, so he took no notice. He had even thought of a rejoinder about searching for an unusual barmaid capable of appreciating the elegance of Maxwell's equations, but decided not to use it and stuck to his line — this problem was an important one and deserved a full solution. So, rather than argue on Rutherford's ground, he picked out the most elegant equation he could find in his treatise and pointed it out to him.

'Look,' he said simply. 'Isn't that beautiful?'

Rutherford, however, was not in a mood for aesthetic considerations and Schott's argument fell on deaf ears. He started thumbing through the pages, looking for a clear conclusion, and was pretty disappointed with the outcome of Schott's work. As he could see without jumping in, the theorist had accumulated a whole series of rather similar examples. However, they all demonstrated that, in any realistic geometry, circulating electrons would necessarily radiate. This was not the result he had been hoping for. It was very unfortunate that Maxwell's theory ran counter to his Planetary Atom. Why beat about the bush?

'All I was asking you for,' he said, 'was a simple and condensed example proving or disproving the possibility of a Planetary Atom. Instead, you have brought me a huge treatise full of equations. This will take me a lot of time to digest, but at the end of the day, I am not really sure it will be worth the effort. I could end up none the wiser and it may well turn out, when I have finished ploughing through all these pages, that it never even answers my initial question. As you know, I am a great believer in the simplicity of things. I am inclined to hang on to broad and straightforward ideas like grim death, and I will do so until the evidence becomes too strong for my tenacity. Why is this so complicated? Surely, Nature doesn't do it this way.'

'Trust me,' answered Schott, 'The full answer you wanted is in here. In fact, you have unearthed one of the most important and fundamental problems in the whole of physics. It was the single crucial example missing from all the work of Maxwell. In retrospect, he should really have thought of it. So, we have added something to his theory. I have been able to uncover, thanks to your initial question, some of the most remarkable properties of the emitted radiation which, by the way, depends only on fundamental constants of Nature. This is all quite new. The equations I have derived are most fascinating and many had never been written down before.'

'This is all very well,' said Rutherford, 'but I am not really interested in the radiation, because I do not believe it exists. I insist that the Planetary Atom is stable, so there must be something wrong in what you have worked out. It is going to be rather hard to put my finger on it in this great heap of paper you have produced. That is why I wanted you to keep your arguments short and simple.'

'The ball is in your court now,' countered Schott. 'At your request, I have given the full theory of this system. Now, if there is something wrong in what I have derived, it can be exposed. I have laid it all before you. Everything is there. Nothing is hidden. In my opinion, what I said to you on the first day is true. Electrons cannot circulate without radiating away all their energy. That is the reality of the situation. If I am wrong, my mistake is in this treatise for anyone to find.'

Rutherford frowned, but he could hardly refuse to study the considerable amount of work Schott had done. That would have been both impolite and ungrateful. He picked up the wodge of paper and assured Schott that he would read it carefully and get back to him. He could not decently do otherwise, but did not relish the prospect of having to read so many pages his colleague had written specifically to answer his original question.

11

A Lady Writes
a Letter

In the week that followed, Ernest Rutherford was rather moody. He did not really like long and complicated mathematical derivations. He withdrew to his office with a lot of paper covered with complex formulae and became very absorbed in what he had to read.

Eileen, who had been told several times not to disturb her father because he was particularly busy, understood immediately that something unusual was happening. Her father seemed less welcoming than usual and less willing to talk, even with her, or to discuss recent events of the day. She sensed something odd about the atmosphere.

Then, she overheard a short exchange between her parents which aroused her suspicion. There had probably been some kind of disagreement in which her father was involved. He said something about the arrogant attitude of mathematicians, which made her fear the worst. Yes, that must be the key to the mystery, and perhaps explained why she had not been told anything about it. Obviously, it must involve George Schott. He had clearly written, said or done something bad which had upset her father. It was obviously quite serious. Normally, Rutherford was willing to listen to contrary opinions and perfectly happy to join in discussions involving opposite points of view.

It sounded very much as though there had been some mistake on the part of George Schott: surely, her father was willing to listen to anything he said and, indeed, had a high respect for his opinions. So, this must certainly be some kind of misunderstanding. Adults were sometimes unexpectedly sensitive about small matters. She had noticed that on a few occasions before. But this was a worrying situation for Eileen. Adolphus was a good friend. She became concerned that, if the misunderstanding persisted, it might turn into a permanent disagreement of some kind. That would be a shame. George Schott might no longer be so welcome in the house if the matter was not soon resolved.

Not seeing her friend any more, when he had become such a regular visitor would be sad. They always spent a few happy moments together in conversation when he came. Being without his visits would be a big disappointment. Certainly, something should be done to sort the situation out before it got any worse. It was very likely that Adolphus was quite unaware of what he had done wrong. Maybe a signal of some kind, a few words of explanation might help. Eileen spent a few days wondering what she might do. There seemed to be no easy solution.

Then, quite suddenly, as she was mulling it all over, she had an inspiration. It was obvious. She should write him a letter. That was definitely the only way. She would explain what she knew and advise him to defuse the situation or resolve the disagreement in one way or another. She was pretty sure Adolphus would get the message. He would heed her advice, because he was a very sensitive person and she trusted the friendship they had for each other. This was obviously the right way forward. If she didn't write such a letter, he might never know what had happened and might never even discover why he was no longer welcome in a house where he had always been very popular.

So, Eileen searched for the right kind of notepaper to send George Schott a message. Of course, it had to be really nice paper. Something very feminine. Naturally, since he was such a sophisticated person, he would expect her to write to him using something refined. Then, she remembered that her mother had just the right kind, a very subtle, lightly coloured letter paper with rough edges and matching envelopes. She had always wondered why her mother kept that kind of paper, since she never wrote letters to anybody, but here was a chance to use it, and surely she would not miss just one sheet and one envelope.

So, Eileen fetched a pen and, in her best and most regular writing, composed the following letter:

My dear Adolphus

I hope you will forgive me for writing. I had no other way to warn you. Something you wrote has upset my father. I don't know why or what it is about. For the past few days, he has been grumpy and is working alone in his room. I can tell he is annoyed. You know how much I enjoy our conversations when you come. I would be very unhappy if you could no longer visit. I am sure you did not mean to offend him. So, please consider some kind of apology.

Otherwise, there would be a problem. You might no longer be able to come. I would be disappointed.

Your faithful friend

Eileen

She read the letter over several times and, on the third reading, decided to add just one word she thought was important: *very* in front of *disappointed*. He would understand.

Then, she sealed the envelope, wrote the address, which she knew, dressed as though she were going for a walk and slipped out of the house to take it to a nearby post office. She was a little flustered, as she had a feeling it was not quite the right thing to do, and her heart beat faster. She knew that decent young ladies don't just write letters for no obvious reason, even to a friend and she was a little worried in case her mother noticed something and she had to give an explanation. Fortunately, her little escapade went unnoticed. It was a beautiful day and the sky seemed amazingly blue as she walked back home. She felt lighter and even slightly elated on her return. She was proud to have taken a courageous step. Maybe she had acted against strict reason, but what she had done was in accordance with her heart.

For the following few days, she watched and waited for the post in case anything should arrive. Suddenly, she realized that there might be an answer to her letter, and how could she explain the sudden arrival of an envelope bearing her name? That might seem odd. She knew the delivery times and watched out, just in case. But there was no reply for quite a long time.

In fact, Eileen's letter reached Aberystwyth very quickly. It was a considerable surprise to George Schott that Eileen had written. He felt embarrassed by it and wondered how to respond. Of course, he knew exactly what the message was about. But what could he do? He did not want to disappoint Eileen and, at the same time, there was no way he could remedy the situation. He understood perfectly well why Ernest Rutherford did not enjoy reading his treatise on circulating charges. It was obvious that his Planetary Model did not hold water according to Maxwell's theory and this was a pretty fundamental issue for him. However, there was nothing George Schott could apologize about. This was just a fact, a mathematical reasoning he had followed through to its ultimate conclusion, not something that could be waved away or hidden from view. Once the truth was out, there was no moving back on it.

On the other hand, Eileen had written a very spontaneous and heartfelt letter. What should he do about that? Somehow, he had to give a reply, if only to show that he had received it and had understood the importance of her message. If he wrote, what would happen to the letter? Writing a letter to such a young girl was a bit awkward. What would the Rutherfords think of him if it were discovered? Their conversations were one thing, but there was nothing secret about them. Everybody knew that they liked each other and that they enjoyed sitting together for a chat whenever he visited. Writing letters was another matter. There was a degree of secrecy in that. It wasn't really the done thing, especially with such a young girl. George Schott spent a few days wondering how to react.

He would naturally have preferred to catch a train and travel all the way up to Manchester. For Schott, that seemed much easier than writing a letter. But, apart from the time and effort needed to get there, there was no point in turning up suddenly if the situation was really as tense as Eileen's letter seemed to suggest. So, really, he could not see much alternative to writing something and posting it to her.

Writing a long letter to a young lady (as he now saw her) was out of the question. It was definitely going a step too far. Anyway, what could he say? It would have been necessary to explain the content of his treatise on circulating charges, which was hardly the right subject for a letter to a young lady, as even Schott realized. Then, he would have had to commit to paper why this might have upset her father, what the arguments were on both sides, etc., etc. No, this was quite unsuitable as a subject for a letter and was actually not the kind of information he could send to a person in the Rutherford household, even though he knew she was a very faithful friend. After all, she was the daughter of the family and it would have put her in an even more delicate position than she already was.

After hesitating for a few days, he hit on an idea which, to his mathematician's mind, seemed to avoid all the pitfalls and to solve all the difficulties he had imagined. He sought out a small sheet of plain white notepaper and, trying hard to attenuate the hard angles of his very strict handwriting, just wrote the following words:

To my very dear Eileen,

thank you

Adolphus

This curt note seemed quite enough to him. It never even occurred to him that she might be disappointed to read so few words. On the contrary, he felt that he had solved a problem for her and that this non-committal reply would protect her in case it went astray. Although George Schott had an appetite for mathematical complexity, he was definitely not gifted at untangling female psychology and really did not know how to deal with writing to a young lady, least of all one who had so plainly laid bare some of her innermost feelings.

Mathematicians, it is often said, tend to be laconic. This was surely true of George Schott. The strangest part of the story is that he was convinced he had answered Eileen's letter. He never realized that, in fact, she was telling him something else through what she had written, perhaps even through the act of writing to him. In a way, he had surely understood that, but he was out of his depth when it came to human feelings and he had really no idea how to respond. This was made all the more difficult by the fact that, somehow, he felt exactly in the same way towards her as she did towards him. But, as often happens, she knew more than he did. Being female, she was actually much more mature and clear-sighted than he was, despite the age difference.

Schott's reply, of course, was a big disappointment to Eileen. This was not at all what she had expected. Somehow, even if it was unrealistic on her part, she had been hoping for a long and beautifully phrased letter with a careful appreciation of all the emotions she had felt when writing to him. How could he possibly have missed what she was trying to communicate? He was a sensitive person.

Why didn't he say more? Why did he not express himself as she knew he could, as he would have if they had been sitting right next to each other?

After his note, Schott's relations with Ernest Rutherford also seemed to become cooler. Maybe, in fact, this was just his impression, but 'the boss', as many called him, still did not respond about the treatise on circulating electrons. The days went by and there was no reply, despite all the work he had invested in putting it all together. Although Schott prided himself on having a very even temperament, he did feel a little hurt. Maybe that is also why he stayed in Aberystwyth longer than usual that autumn and did not venture as far as Manchester. Somehow, he had the feeling that he no longer quite belonged in the intimate and exclusive circle of Rutherford's scientific friends.

Chapter

12

A Danish Theorist
Joins the Fray

In March 1912, the young theoretical physicist Niels Bohr left Cambridge, where he had come to complete his studies. He had decided to go to Manchester, because that was where the world-famous physicist Ernest Rutherford was working.

Schott, who had finally decided to visit more or less at the same time, was a little upset to discover that Rutherford was transferring all his attention to the activities of the young Danish theorist and seemed hardly interested in what he was saying. Had he really studied the treatise Schott had written for him? He said so, but his attitude was very unclear. At any rate, he did not seem very keen at all to discuss the matter further. Again, Schott was disappointed. He had worked so hard on his refutation of the Planetary Atom model that it was hard to accept such indifference on the part of the person he had written it for.

Worse still, this young man, Niels Bohr, seemed to have captured Rutherford's attention, despite the statements the 'boss' always made about not believing in highfalutin' theories. Schott could not help feeling that Rutherford was somewhat susceptible to flattery. No doubt, the young Niels Bohr (what a name!) had ingratiated himself by going along with everything the professor was saying about the somewhat absurd properties of his Planetary Atom. There was something of the ageing monarch about Ernest Rutherford, and perhaps his own importance was going to his head. This young man from Denmark obviously knew how to manipulate an old professor. Agreeing with everything he said was the worst way to make real progress in science.

This was an unfortunate development. Proper scientific discussion was really needed. To restore sanity, George Schott determined that he would bring up the questions raised by Paul Ehrenfest, the German theorist and confront Rutherford with a different opinion.

'You know,' said Rutherford when Schott eventually managed to bring up the subject of the treatise with him, 'your mathematical deductions are indeed very interesting, but they are rather complicated and I am a simple person. I don't see how theorems, elegant though they may seem, can stop the atom from being as it is. I tell you it is Planetary and I don't need to tell you, of all people, that it has to be stable. What more do you want? If you wish to prove otherwise, I think you will need to turn yourself into an experimenter, or at least to explain to me what measurements might confirm your opinion.'

Rutherford was gradually becoming the pope or even the prophet of a new subject he was creating, called nuclear physics. He was beginning to feel that, if theorists could not confirm what he had to say, it was their problem and not his. Schott began to suspect that he had probably not worked through the full detail of the great treatise he had written on radiation emitted by circulating electrons. No doubt, he had overestimated Rutherford's ability to follow long mathematical derivations. Perhaps it was true that he should have kept things simple, or at least provided some kind of summary of his conclusions. Obviously, Rutherford had lost his way somewhere in the middle of what he had written.

Since he was not making progress with the one person he had tried hardest to convince, George Schott decided to communicate his work to others around him, some of whom took a greater interest in theory, as well as former friends and colleagues in Cambridge.

Many of them had accepted Rutherford's model of the atom but, even among the physicists who did not share Schott's opinion about the structure of the atom, there was universal admiration for the mathematical quality of his work and the elegance of his proofs. Schott's treatise was, in fact, the very first complete study of the electromagnetic radiation emitted by electrons on a circular orbit of constant radius. What emerged was that, in order to maintain the electron on this unstable path, it was necessary to supply it continuously with an amount of energy equal to the energy lost by radiation. For Schott and for several other contemporary physicists, this was the death knell of Rutherford's Planetary Atom.

Henceforth, it would be recognized (and had now been most definitely established) that this model was inconsistent with Maxwell's theory of electromagnetism.

Rutherford had no desire to accept this conclusion. In fact, he understood Schott's position perfectly well, but he had no intention of revising his model. He was also aware that there was a distinct chance he might be wrong. After all, Maxwell's theory was a pretty robust leg of classical physics. Nobody had done any better and it would have been pretty risky to come out saying that it was just wrong. So, Rutherford still stuck to his guns, but avoided battling in public, while Schott, who was naturally reserved and hated polemical debates, merely circulated his treatise among friends. Rutherford behaved as though he had not studied Schott's work in complete detail. In fact, this was not true. He had worked through it line by line and just pretended he had not grasped the obvious conclusion. He knew exactly what Schott had deduced and was forced to conclude that, in some way, perhaps he might seem right. But he had no better idea to propose to account for his observations. This was his dilemma. Surely, nobody could disagree with electromagnetism. So, for the moment at least, Rutherford and Schott had to agree to disagree and leave it at that. Yes, Rutherford was often a bit of an iconoclast, but not enough to challenge the mighty Maxwell.

Two opposite camps were beginning to form. Rutherford was forced to recognize, not only that Schott had joined the 'other camp' namely the opponents of his model, but also, in view of the power of his intellect, that he would undoubtedly become its general. This was clear, because George Schott was unquestionably the leading exponent of electromagnetism in the whole country. Nobody else had such a complete grasp of all the subtleties of the subject.

Rutherford was obliged to accept that his strategy had not achieved the desired result. So, he became a little bit more cautious in his defence of the Planetary Atom. He just stuck to the observational facts and avoided any mention of the theoretical implications, especially in public presentations. Indeed, he became very cautious about this thorny issue. As he was looking for support of his own point of view, he communicated some of Schott's conclusions to the young Danish theorist Niels Bohr and, as he did so, revealed that Schott would probably turn out to be the greatest and most radical opponent of his new model. He knew that Bohr was strongly attracted to his own ideas and Rutherford felt it was probably important that he should be aware of the formidable opposition he was facing.

Having thus ignited a long war between opposing factions of theorists, Rutherford retired to his fortress of experimental physics, from which he could safely observe the scene without becoming too compromised in what was to follow. He was not one to mix opinions with personal relations and his admiration for Schott was in no way diminished by their disagreement. Perhaps even the reverse. At this point, he was really rather relieved to have handed over the theory of the planetary model to someone else. Mathematical physics was not really his *forte* and it was better to leave it to the experts.

He was so relieved, in fact, that he suddenly remembered he had been responsible for starting George Schott on the long path he had followed to put his treatise together. At last, he thought of thanking him for the enormous amount of work he had done and he renewed words of friendship he had somewhat forgotten during their period of intense argument. In fact, he suddenly realized how badly he had treated

George Schott and became aware that he needed to personally make up for having let a professional disagreement get the better of their friendship.

He realized that Schott had not been round to visit him for some time and made a point of inviting him home again. There was nothing artificial about it. Rutherford was always very direct. Handing over Schott's papers to someone else removed a cloud under which Rutherford had worked for some time. Now that the cloud was gone, he felt free again and could see their story for what it was: an essential step in the development of physics.

However, Rutherford did not only have admiring friends. In Cambridge University, in particular, quite a few ambitious characters had observed his rising fame with concern. They suspected that his ultimate aim was to return there and claim a high position. They saw his increasing importance as a threat to their own careers and were beginning to club together around the objective of keeping him out. They had developed a phobia for this noisy New Zealander and his rustic ways, for his loud voice and permanent dominance in the pages of scientific journals. In fact, they were becoming quite worried about him.

For the anti-Rutherford faction in Cambridge, Schott's treatise on the radiation emitted by circulating electrons was a wonderful opportunity. If Schott was right (and how could it be otherwise?) here at last was the chance to shoot down the man who was setting himself up as the father of a new branch of physics. Schott's writings were not only very impressive and firmly grounded. They also avoided naming anybody in particular and concentrated on ideas. They could not be labelled as partisan. This was exactly the right approach. For real experts who could follow all the mathematical detail, here was the ultimate proof that Rutherford was wrong in his conclusions.

As soon as Schott's work began to circulate, a conspiracy was formed. A plan was born to celebrate the achievements of the former student of Trinity College, the distinguished mathematician from Aberystwyth. There was some consultation about how to draw attention to his treatise and give it the importance it deserved.

After some debate, it was resolved to attribute the Adams Prize to George Schott. Even if few people were aware of this distinction outside Cambridge, it was in fact extremely prestigious and well deserved for a number of reasons. The prize had been created to celebrate the work of John Adams, the rival of the French astronomer Le Verrier who had somehow managed to convince the world he had discovered the planet Neptune, whereas in fact, the British discoverer had been overlooked. Among previous recipients of the prize (and this was perhaps the most important point) was James Clerk Maxwell himself. What could be more appropriate? This was exactly the right accolade for George Adolphus Schott. It was also a statement and would be understood by all the real experts.

Furthermore, the prize brought with it a unique opportunity to spread the word. Schott's treatise was published by the most prestigious of all scientific publishers, the Cambridge University Press. This was indeed recognition. To be published by those who had first printed the works of Isaac Newton was something to be taken seriously. And of course, on the title page, the Adams prize was announced as having been awarded to the author.

Rutherford received several copies of the book. One was a friendly gesture, with no malice intended, from Schott, who was genuinely grateful to him for having brought up such a fascinating problem and wrote him a little note on the title page to say so. Others, perhaps, were not so friendly, from those who were firing a warning salvo across the bows of the ambitious New Zealander.

Of course, Schott's work had other consequences. It was also bad news for all the French supporters of Jean Perrin's atom with electrons also revolving around the nucleus like little planets around the sun. Clearly, there was exactly the same difficulty. Rutherford was curious about the reaction there would be in Paris, so he sent a copy of the book to Marie Curie with a little non-committal note pointing out that there seemed to be a difficulty (according to the author of the treatise) when circulating charges were present.

However, Marie Curie was already very ill. She had neither the time nor the energy to start reading a long book full of mathematical derivations which were admittedly very elegant but remained rather complex. So, she donated it to the library of the University of Paris, which seemed to be the right place to keep it. Oddly enough, despite the prestigious publishing house and the mention of the Adams prize on the title page, the book did not attract much attention in Paris. It led a solitary existence on the shelves. Some potential readers would lift it up from time to time but, finding it rather heavy, would quickly put it back in the place they had found it. Truth be told, it seemed interesting enough, but was too thick and too voluminous. Gradually, it was moved to a neglected corner alongside similar works of inconvenient format. Researchers had ceased to take much interest in the subject as it seemed to have been resolved in another way. There are quite a number of works in scientific libraries which fall under this category.

The original edition of Schott's book had only a modest printing run. The publisher knew from the start that he would not sell many copies of such a specialized work. That is why rather few copies of the original edition have survived to this day. One of them, in the British Library, was even lost when the stock was moved to a safe place outside London during the bombings of the Second World War. The book was only found and restored to the shelves some forty years later. But the truth is that nobody missed it in the meantime.

For some reason, Schott's book was only read by a rather small number of people. That was very unfortunate as it was really his life's work and he was too discrete to advertise how much he had done. Even in Great Britain, where he worked all his life, the subject of atomic models simply moved on along a somewhat different path.

Why so little interest in his work? The reason was perhaps surprising. The young theorist Niels Bohr had taken absolutely no notice of Maxwell's theory or of Schott's careful arguments. He just forged ahead without the slightest compunction and asserted (with no proof at all) that the atom had stationary states and that these did not radiate. This was regarded by him as a postulate. There were new laws of physics. This brazen approach seemed to cast aside all of Maxwell, which should have shocked British scientists even more than the others, especially in Cambridge. It would probably have been considered as the work of an impertinent young theorist if it had not led to an impressive series of deductions which fitted very well with observation. In fact, Bohr had just invented a new kind of physics, which he termed 'microscopic'. He claimed that it followed completely new rules, which were strange and counter-intuitive. It was a radical step in the direction of what we call quantum mechanics today, but the step was so radical that it left many physicists rather confused about his proposals.

Schott was not the only scientist to be disgusted by such a brutal approach to a complex and important problem. Even Rutherford was astonished by such blatant arrogance on the part of a young theorist. He was also (it has to be said) a little peeved: he discovered on reading Bohr's papers, a hypothesis he could very well have put forward himself. All he should have done was to come out into the open and assert plainly what he had always believed: the Planetary Atom has states that do not radiate. The electrons just go round forever without losing energy. He had always told Schott this was the case. He could easily have introduced Planck's constant himself into the problem. He had all the facts available.

The man who had stopped him in his tracks was George Schott. Why had he listened to his objections? Schott had been awkward from the very beginning. He had brought up all kinds of counter-arguments based on his knowledge of electromagnetism, and Rutherford had bowed to his superior understanding of Maxwell's theory. But in the end, this had all been wrong. Rutherford felt betrayed. He should have followed his own instinct and just taken no notice of Maxwell or of Schott.

This could have been his greatest discovery. It could have been called 'the Rutherford Atom'. Instead, it would be known as 'the Rutherford-Bohr Atom'. But Rutherford knew very well that words get shortened almost immediately by those who utter them. This was really too much of a mouthful. Sooner or later, it was inevitable that it should just become 'the Bohr Atom'. So much shorter. And in a way, more accurate, because

Rutherford had missed the so important step his work was leading to. And it was all because of Schott. It was his fault.

There was a strange paradox in this situation. Rutherford had always warned against too much respect for theorists and their idle speculations. And it was Schott, with objections grounded only in theory, who had made him miss the boat at the last minute. In the end, it was he, Rutherford, who had been too respectful of theory. A doctrinaire mathematician had made him miss his chance. Rutherford, the true father of this new physics, could not help feeling aggrieved. He might have led the revolution from the front and someone else had now stepped in to do it in his place.

On top of that, Schott was now leading the opposing forces in the debate. The polemics around the Planetary Atom had really caused a lot of grief on both sides.

Quite by chance (but maybe chance is not the right word as it seems to introduce something too random in the context), Schott found he was walking down a corridor at Manchester University and Niels Bohr had suddenly appeared at the other end, walking towards him. George Schott hated arguments. His first instinct was to change direction, but it was too late for that. He would have given the impression he was running away from a confrontation. Even worse, in Schott's eyes, it would have seemed impolite towards a colleague. So, he just paused and wondered what to do.

Niels Bohr, on the contrary, saw an opportunity in the meeting and rushed forward towards him to shake hands:

'What a pleasure to meet you, Mr Schott!' he said, 'I was meaning to get in touch and congratulate you on your beautiful book. I wanted to tell you how much I admire your work. The boss gave it to me to read. It is a monumental piece of mathematical physics. Very impressive! And despite what other people may say, believe me, I agree entirely with your conclusions.'

'Really?' said Schott, 'somewhat staggered by this declaration. Did you also say that to Professor Rutherford? And what was his reaction?'

'Of course I did,' answered Bohr, 'I told him exactly what I have just told you. But Professor Rutherford is an experimenter, not a theorist. In fact, he has rather little understanding of higher mathematics. That is why he was not really able to make an informed judgment on your treatise. I explained to him that everything you have worked out and all your conclusions are completely correct, absolutely pertinent and remarkably profound. The only point at issue is whether they apply to atoms, and I believe that they do not.'

'What do you mean by that?' asked Schott. 'Why do you think atoms should be such a special case?'

'They are not built on the same principles,' explained Bohr. 'If you like, it is because they do not have the same creator. If we were dealing with a man-made system it would be different. If a physicist were to try and create an atom, he would do it exactly in the way you describe. He would start with a free electron and guide it into a circular orbit around a positively-charged nucleus. Maybe one day, such experiments might become possible, although the idea may seem far-fetched. In this scenario, he would necessarily witness all the effects you describe in your book, beginning with the orbital radiation you discuss in such detail. This is what I call "the classical limit".'

'Very well,' said Schott, 'but then the electron in this circular orbit will gradually lose energy and fall into the nucleus. No other outcome is possible.'

'Not at all!' objected Bohr. 'Your problem is that you are trying to deduce the properties of the atom by using the classical model. The atom is not a classical object. It is true that, in the classical limit, everything happens exactly as you say. But the mysterious behaviour of atoms only appears in the opposite limit, where the system becomes so small that, in fact, it is just unable to radiate. This is such a tiny system that the underlying physics becomes quite different. I call this "the microscopic limit", where atoms no longer radiate and so become stable. Microscopic physics is just completely different. To put it simply, it is counter-intuitive. We have to learn new rules.'

'That is a bit like the Indian rope-trick,' answered Schott. 'If the structure of the atom remained planetary in that limit, I am afraid it could not be stable. It would have to continue loosing energy. You can turn physics on its head by making new postulates, but real atoms must still radiate, probably by very mysterious processes, more subtle than we can even imagine. Anyway, we know that spectral lines of fixed energy are emitted. That is not consistent either with nearly circular orbits or with a continuous loss of energy. According to your ideas (which anyway are only suppositions) the atom would be both planetary and stable in its lowest energy state. I beg to differ, because all these contradictory requirements you introduce are simply not consistent with Maxwell's theory.'

Bohr smiled.

'I don't in any way claim to have solved all the mysteries of atomic structure,' he answered. 'All I am saying is that I have found a clear relationship between all the observations made so far for hydrogen and that they all sit comfortably within the planetary model. After all, this is what theory should do: it should bring all the actual observations under one roof and adapt perfectly to experiment. No more and no less. Yes, it is true that I have to sacrifice some of the previous rules of physics to get there, but that is part of the game. All I need to assume is that Maxwell's theory, for some reason I don't claim to understand, does not apply to very small systems like atoms. In particular, we just suspend the rule that accelerated charges necessarily emit radiation. The laws of physics must obviously depend on the size of the system. That is all we need to say.'

'That is very cavalier,' objected Schott. 'In order to explain a few observations for atoms, whose structure we do not even understand, you ride roughshod over centuries of well-established science. You don't only throw Faraday and Maxwell out of the window. You destroy the proper foundations of mechanics, due to Hamilton and Lagrange, even going back to Newton...'

'I gave you my opinion,' said Bohr, 'I agree with you that the new rules don't fall into place easily. We must get used to them because our intuition derives from macroscopic, not microscopic systems. So, when the rules are not the same, they become unfamiliar to us. That is the whole difficulty of understanding microscopic physics. However, I do assure you, it is the right path. The systems you describe are also completely correct. They do exist, but in another limit. They are the large systems. Maybe they exist as well in Nature. It is just that we have not found them yet.'

'You are making a very strange distinction between large and small,' Schott remarked, 'it doesn't make sense at all to me.'

'The fact is,' continued Bohr, 'just as man cannot manufacture atoms, he would find it equally difficult to make the giant atoms you describe. We are talking about systems created by Nature, not by man, and we do not decide what properties they should have. We adapt to the ones we discover. It would be very difficult to make your giant atoms. You would need to accelerate an electron all the time in order to maintain it on an orbit of a fixed radius and allow it to recover the energy it loses by radiation. That would be pretty difficult to achieve and would require huge amounts of energy. Making systems is much more difficult than breaking them. That is just life. Rutherford, having discovered what he calls the nucleus, is now busy trying to split it apart...'

'On this last point,' conceded Schott, 'I entirely agree with you. He reminds me of small children who discover a new toy and then must break it.'

On all other matters, they had opposite points of view, so Schott, although in some ways he was happy to have cleared the air with Bohr by the discussion, remained on his guard concerning the atomic model.

I really can't understand, he thought to himself afterwards, *why the laws of physics should be so different for systems of different size. This mysterious constant of Planck which Bohr introduces into his atom originated from the properties of radiation but, in the same breath, he wants to prevent the atom from radiating when it really should do. None of this is very logical. He just decides to turn off Maxwell's theory in a completely arbitrary way. Why should the properties of free electrons, which do radiate, be so different from those of the electron when it is bound to the nucleus of an atom?*

Schott belonged to the small number of people who can never rest satisfied if they do not achieve a complete understanding of a problem and who reject all forms of approximation. The idea that an area of mystery might persist which would prevent one from following all the properties of a physical system under any circumstance and analyzing its evolution in the smallest detail was very irksome. He could hardly live with it.

After his discussion with Niels Bohr, Schott was reminded of a remark he had heard from the French physicist Louis de Broglie who, like him, was a great admirer of Albert Einstein. De Broglie had tried to understand the properties of the electron by associating its motion with a wave. Then, he had tried to justify the existence of stationary states by comparing them with resonances in an organ pipe. That at least had some virtue. Schott did not believe in the analogy, but it had an element of plausibility in it which was not completely arbitrary. So, Louis de Broglie, at least, was a real scientist. He did not resort to vagueness. And he had said:

'Niels Bohr is such a master of *chiaroscuro* that we ought to call him the Rembrandt of modern physics.'

Yes, Louis de Broglie was completely right, thought Schott. *If the new physics they keep talking about is just based on reasoning of this kind, surely it is a fashion and, like all fashions, will pass.* He tried to imagine what it would be like for Niels Bohr in his old age attempting to explain to new generations of young scientists his odd conception of the properties of atoms and the strange model he had concocted in order to try and relate their properties to each other.

Once again, this is 'Phlogiston,' he thought. *There is something of a new religion in this so-called 'new physics.'*

If somebody had asked George Schott whether he believed in God, or perhaps even in the Devil, he would not really have been able to produce a convincing answer. He was not religious in any sense of the word. And yet, at the same time, he did not like the idea that nature should contain anything irrational or impossible to explain. Somehow, he felt that man had been given a brain in order to figure out causes and understand certain mysteries of the universe even if he could never achieve a full and complete picture. It was just a question of correct reasoning. In this respect, he was very close to the deist philosophers of the eighteenth century, people like Voltaire who always argued, when faced with the question: 'Does God Exist?' that one cannot imagine a clock without a clock-smith.

So, the idea that there might be a domain of science called 'microscopic physics', with different rules from another domain more familiar to us, was very strange indeed. How did one get from one to the other? And what about the intermediate range? Why should there be a frontier between the two and what did that really depend on? All of this seemed quite obscure to him. On the subject of the existence of God, George Schott would not have found anything much to say, but on the question of any kind of incoherence in the midst of the universe, in his view, there could only be one reply. It was strictly impossible. From the scientific point of view, this would be something analogous to blasphemy for a theologian. Bohr wanted to introduce vagueness and uncertainty. These concepts were anathema to Schott. He was broad-minded in many other respects, but this was really not consistent with the proper pursuit of science.

Not understanding something, in his opinion, was not necessarily a mark of ignorance. This was the original state of mankind but great strides had been made, so it could not be considered as inevitable. Schott's mind was open to all kinds of new ideas, except the notion of some kind of inherent fuzziness of nature which would hide away cogs and wheels and prevent a full interpretation. So, Bohr's approach disturbed him greatly. In fact, he found it quite unacceptable.

Since he was gifted with great mathematical insight, he set off on a different course. He would search for all the solutions of Maxwell's equations which might yield stable spherical charge distributions capable of rotating without emitting radiation. If there were any, he would find them. He had learned that this might be possible by reading a paper by the German physicist Paul Ehrenfest. This seemed to be a good idea and nobody else was doing it, so Schott decided he would follow this path.

It was a promising direction and he discovered several interesting configurations as soon as he started to work in earnest on the idea. Unfortunately, none of them seemed to fit the observed properties of atoms. Obviously, the solutions would be rather complicated, as Rutherford had surmised.

In his dialogues with Nature, Schott always hoped he could find complete and perfect order, rather like French gardeners who seek perfect symmetry in plants. He was against, not only randomness, but also anything lopsided. Nature, however, did not seem willing to submit to all his intentions and, on this occasion, seemed quite determined to stop him going any further.

Ultimately, a scientific theory has something to do with one's philosophy of the world, just like poetry. For George Schott and for Louis de Broglie, there should be nothing vague or disorganized. This was also a matter of choice.

13

Schott Slips through
the Great War

The Great War brought everything to do with the Planetary Atom to a sudden halt. This was yet a different answer to the question Eileen had raised in her letter. Would relations between George Schott and the Rutherford family, in the end, be broken off? It turned out that nothing occurred quite as severe as she had feared. There was no real conflict around the Planetary Atom, simply because the world was busy fighting and the activities of all laboratories were diverted towards what was described as 'The War Effort'. In one way or another, every citizen was expected to become involved. Conscientious objectors were to carry stretchers if they did not want to carry guns. Women would become nurses, and intellectuals, however abstruse their normal work, had to find ways to make themselves useful.

Schott no longer had any reasonable excuse to travel to Manchester as he had grown accustomed. Even the 'boss', as everybody now called Rutherford, had to put everything down and work on something deemed more useful. He was not too happy about the intrusion of world events into his daily life and said so pretty clearly. On one occasion, when he arrived late for a meeting to discuss submarine weaponry, he was reprimanded by the chairman of the committee and gave the following response:

'Gentlemen, I apologize for being late but, in the meantime, I have disintegrated the nucleus of the atom. Believe me, that is a more important event than your war.'

Nevertheless, even Rutherford had to toe the line and follow orders. Apart from Marie Curie, who participated heroically close to battle lines in her legendary truck kitted out with an X-ray tube, diagnosing shrapnel wounds and broken bones, most physicists were commandeered by their respective nations to invent various kinds of new weaponry. Henceforth, the purpose of their research was to be the invention of new ways of killing people. There were a few exceptions, notably amongst the younger scientists.

A young and brilliant theorist, Ralf Fowler, from Rutherford's research team, and a young physicist, Patrick Blackett, saw active service, the first in the Infantry and the second in the Royal Navy. Eileen, who knew them both, was horrified at the idea of all the dangers which awaited these brave young men.

Rutherford, who had a more traditionalist view of the need to serve King and Country, was not too upset by their departure. He had already decided he wanted to recruit Niels Bohr as the leading theorist in his group, so Fowler's departure did not worry him. In his view, Bohr was much more important. He was the one who had put a stop to all the useless discussions about the effects of radiation cooked up by Schott. So Bohr was now the hero of the Planetary Atom, and that would be the future when the war would be over.

For the moment, Rutherford was obligated to serve the nation by applying his unequalled experience of useful equipment to the design of military hardware. He was required to invent new methods of communication and detection making use of electromagnetic waves, as in his early years in New Zealand. They had become strategically important in modern warfare. The Russians had already been taught an expensive lesson by the Japanese at the Battle of Tsushima, 1905. Confident in the superior firepower of their ships, the Russians had sent a huge force into battle without any modern means of communication. Very cleverly, on the other hand, the Japanese were equipped with excellent electronic devices supplied by the new Shimadzu company, already at the forefront of technology at the time. The result was a humiliating and disastrous defeat for the Russians.

The young physicist Patrick Blackett was on the deck of a battleship at the Battle of Jutland in 1916. He observed that the British gunfire was missing its targets by two hundred and fifty yards to one side of the German ships and immediately understood that the telemeters supplied by the

Admiralty had not taken into account the change in sign of the Coriolis force at the equator. He recalculated the correction needed on the spot. The guns were readjusted before firing again and the British fleet was victorious.

The Great War was one of the first conflicts in which technological advances were shown to make a huge difference. Progress in science was now looked at with real interest by military commanders everywhere in the world.

However, for some researchers, the war also caused long periods of inactivity which, if at all possible, they did their best to fill usefully. It was under enemy fire in the trenches that another former student of Trinity College Cambridge, D'Arcy Thompson, composed his immortal classic *On Growth and Form* thus laying the basis of a whole new area of science: mathematical biology. However, apart from this masterpiece written under the most improbable circumstances, pure research as such ground to a halt for the duration of the war years.

In times of conflict, reality takes over from inspiration and imposes unpleasant obligations on researchers such as the invention of tools of destruction. In the times of peace which follow, public opinion becomes retrospectively shocked by the horrors of warfare and blames the scientists for their inventions. Rather than assume responsibility for having forced them into such work, it finds some relief from its own guilt in this way and sometimes goes so far as to consider science itself as the source of evil. As Shakespeare observes, men are happy to blame the stars and the planets for their own depravity and the modern version of that is to lay the blame on technology rather than on the use made of it. The natural state of science, left to its own devices, is in fact to be driven only by curiosity, but this is soon forgotten when there is an enemy to defeat.

George Adolphus Schott was fortunate enough to escape such issues. As a mathematical physicist in an obscure place, Aberystwyth, he was regarded as pretty useless and was not even asked to join in designing weaponry. His only fear when he enlisted as a soldier was that he might be required to hold a rifle and shoot at Germans. Since he was something of a German himself, that would have been difficult for him to do. As to fixing bayonets and jumping into trenches in France for hand to hand combat, he had not the remotest idea how civilized human beings could be forced into such extremities of behaviour.

Luckily for him, the sergeant in charge of sorting out where to send incoming recruits immediately identified him as very poor material for soldiering. For a start, he was obviously in bad health and was soon diagnosed as having some kind of recurring but unexplainable form of bronchitis, possibly contagious. He would probably not have lasted more than a few weeks in insalubrious trenches and would have had to be evacuated pretty quickly — hardly worth even sending him there. He was also identified as having an extremely slow response time, poor coordination and had obviously never practised any sport. Finally, he turned out to be some sort of intellectual. The sergeant had dealt with such people before and knew that, through no fault of their own, they make extremely poor soldiers. They tend to misinterpret even the simplest orders, ask irrelevant questions and create confusion by trying to understand what is going on, which properly trained soldiers never attempt. The sergeant was half tempted to send him straight back home when it emerged that Mr. Schott was good at mathematics. Apparently, he had even studied the subject to quite a high level.

This is where the recruiting officer showed just how resourceful he was. George Adolphus Schott was immediately posted in a support capacity as a training officer in ballistics to an artillery regiment. Schott was actually rather happy that a useful task had been discovered for him.

Being sent home in wartime was a very visible and somewhat humiliating experience. He wanted to do his bit like all the others. So, he took up the task with his usual diligence, learned what he was supposed to teach very quickly and was soon explaining to all the soldiers the rudiments of ballistics and how to point heavy guns in the right way, according to the handbooks issued by the War Office. Indeed, he turned out to be so good at this rather uninspiring and desperately repetitive job that he was soon commended and promoted for the excellence of his teaching and reached the exalted rank of lieutenant.

The war where he was turned out to be very boring and the mathematics he had to deal with were so trivial they bordered on extreme stupidity. However, as he told himself from time to time, somebody had to do it or the guns would never be pointed in the right direction at the enemy. So, he did his best in the service of his country, while realizing fully that he might have been doing the same job on the opposite side if his parents had not emigrated from Frankfurt. Luckily for him, he was never actually required to fire a gun himself. The military had very properly separated theory from experiment. His task was to teach the basic principles, and nobody expected anything else from him.

One day, somebody, on hearing that he was good at mathematics, challenged him to a game of chess. This was the beginning of a new life for Lieutenant Schott. After he had defeated all the soldiers and non-commissioned officers, he moved up to playing more senior ranks and became quite a celebrity along The Front. People in the know began to ask about Lieutenant Schott. How come there was such a brilliant chess-player in the regiment? Where had he trained and who was he really in civilian life? It thus emerged in the open that Lieutenant Schott, in peacetime, was actually the captain of the chess team of Cambridge University. Even generals felt honoured to play against him and were no longer surprised to be beaten hollow. Indeed, they rather expected it. To them, it became an embarrassment that a man who was a true captain in civilian life should be a mere lieutenant in wartime. So he was promoted again, this time to a military rank he fully deserved, in line with his standing as the best chess player along a whole sector of the British army in France. Perhaps, had hostilities lasted a little longer, he might have risen to a still more distinguished rank by defeating a few more generals or perhaps even a field-marshall, but the opportunity never arose. So, when the Great War suddenly ended, the burgeoning military career of George Adolphus Schott, Captain of Ballistics at the blackboard and supreme strategist at the chessboard, came to a sudden end.

After the Armistice, he returned to Aberystwyth to resume university duties. In fact, he had pretty well forgotten this interruption in his life of study when a young artillery officer appeared at the porter's lodge and requested to see his former comrade in arms Captain George Schott.

He did remember vaguely having taught ballistics to young Lieutenant Nicholas Fortescue, Bart., whose rather pretentious visiting card flaunted his military rank. Indeed, he even had some recollection of having played chess against him and having found him reasonably able. So, Schott felt obliged to see him.

The young soldier, better known as 'Nick' at the army, had indeed served at the Battle of Arras under the command of General Allenby. Since demobilization, he had run into many problems. Basically, he had no training or higher education of any kind and, in spite of this, no desire either to work with his hands. He felt that the war had interrupted his education and that it was now too late to start studying again. Civilian life had proved very trying for him and there was no recognition for all the sacrifices of the war years.

At first, George Schott thought this was simply a courtesy visit, but soon it turned into something rather different. In fact, the young Lieutenant had come all the way to Aberystwyth to make a most unusual proposition.

'I need a mathematician,' he said. 'Not just any run of the mill mathematician, but a great mathematician like you. If we could work together, I am sure we would soon make a fortune. Let me explain how. I have managed to borrow quite a large sum of money from the bank. Of course, I will need to pay it back, but I have an excellent plan. I have studied the housing market. Did you realize that, during the war, property prices had plunged? In fact, they have never been so low in living memory if you compare them to the price of a loaf of bread. That, you might say, is perfectly normal. So many people have been killed. There is no demand. In fact, many fine houses are standing empty. This is the time to buy. Soon, the economy will start up again, not just in England, but all over the world. Houses you can buy for a song today will be worth a small fortune tomorrow. That is my plan. Invest in property.'

'Since you know that already,' said Schott puzzled by the strange turn of the conversation, 'why are you looking for a mathematician?'

He had worked out that Nick Fortescue had an enterprising character and was no fool. What surprised him was such a strong ambition to become rich without doing any real work, just by playing with money borrowed from the bank. The young lieutenant was amazingly self-confident.

'My problem,' explained Fortescue, 'is that I need to wait. I am absolutely certain that my investments will prosper. There is no doubt about that. In fact, prices are beginning to rise already. In the meantime, however, I need to make a living. I must also pay the interest on my loans. I will probably need to hold the fort for five or ten years before

I become a millionaire. The houses are wonderful and their locations superb, but their value has not risen yet.'

'Surely you can rent them,' suggested Schott. 'The property is yours for the moment. That would bring in a return in the meantime.'

'True in principle,' responded Fortescue, 'but if I fill them with tenants, they lose value. Empty houses are worth a lot more. So, I don't want to do that if I can avoid it.'

'In that case,' said Schott, 'you will have to find some kind of well-paid job to survive.'

Schott had never encountered the kind of situation Lieutenant Fortescue was describing. His own family had a factory and his parents, who had built up the family fortune over the years, had always been there to support him in the background. He had been able to study as much as he pleased and now had a secure position in the university. Although he was not interested in acquiring wealth, he suddenly realized how difficult life must be for a young man like Fortescue who was attempting to live by his wits alone. This was an aspect of ordinary life Schott had never faced. Like so many other people, Fortescue's great ambition was simply to live without actually having to work. During the conversation, Schott began to wonder how he would have managed himself if he had been born poor. On reflection, he suddenly realized that he, also, deprived of means of his own, would have been unemployable outside a university.

'Your situation sounds a bit perilous to me, he said. I am surprised the bank lent you so much money in one go. Surely, you need some regular income to keep going. Are you sure of what you are doing?'

'I have the perfect answer,' said Fortescue. 'The Stock Exchange. In principle, this is more complicated, but I have been studying movements of shares for some time now and, of course, there is also a general rise, just like what will happen for houses. However, it starts earlier. By studying the trends, one can control the parameters, understand the variations and construct predictive models. That is the key. Let me show you what I have been achieved so far.'

And he took out a few neatly folded sheets of paper from his pocket, covered in numbers and hand-drawn curves.

'How did you obtain all this information?' said Schott after briefly studying the pages. 'This must have taken you quite a lot of time to work out.'

'Very simple,' said Fortescue. 'During the war, I gained access to the officers' mess and I was able to study daily movements of shares in the newspapers. I kept it up ever since. I was already interested in markets and I had some idea I could find work in this kind of business. To put it in simple terms, it is a bit like playing chess. You have to guess what is going to happen before it happens. That's all. Notice that every one of the curves shows the effect of the war on the markets. And now, most of them are heading upwards again, except of course the ones for suppliers of military hardware. They go down. But eventually, shares will all recover and resume the slopes of growth they had before the war.'

'Not all of them, surely,' said Schott. 'Some might behave quite differently.'

'A few, perhaps. But the vast majority will do as I say. That is a statistical certainty. Look! Some are already recovering. Very soon, they will have reached the position they were heading for before the war started. That is absolutely inevitable. One can be confident of that.'

'I am not so sure,' said Schott, 'the future always seems uncertain to me.'

'There could be a few fluctuations,' conceded Fortescue. 'They are just due to interference between parameters. That, in fact, is why I need a mathematician. The problem is to extrapolate the curves towards the future and to take interferences between them into account. In principle, it is not very difficult, but it needs a good mathematician to write equations, make sure there are no mistakes. He is the one to figure out that everything has been properly taken into account.'

At first sight, this seemed a new kind of challenge to Schott. Despite his scepticism, he could not help asking himself whether, perhaps, Fortescue might be right. Was it possible to predict something about the evolution of the markets? It seemed rather far-fetched, but mathematics has many hidden powers. Maybe there was something in the idea. Even so, there was still something a little unsavoury about the whole concept. For Schott, the business of turning knowledge into wealth seemed truly vulgar. He remembered being taunted at a college dinner by someone who hated mathematics and who had said:

'If you are all so clever, why aren't you rich?'

There was something unethical, in his opinion, about the notion of turning the pure beauty of equations into pounds, shillings and pence. The idea had not yet occurred to him and was a bit frightening. But, on the other hand, if this kind of alchemy was indeed possible, then it was perhaps wiser to be aware of it. Hidden powers of this kind might have a lot of bad consequences. This was something to be wary of, if at all true.

'Your second name is already Adolphus,' joked Fortescue, who had somehow discovered his hidden German identity. 'Just imagine what you could do with your amazing talent in mathematics. Don't forget that knowledge is power. If we join forces, you may become the Doctor Faustus of our century. All you need to do is apply some kind of algorithm to find the right combination of companies, the ones whose shares are surest to rise in value during the next trading session. It is such an obvious way to become rich that, if I were in your shoes, I would even invest my own money into the scheme.'

Schott was rather taken aback, but willing to listen:

'Leave me your curves for a day or two,' he said. 'I don't really believe one can predict the future, but I am willing to examine the problem. I will try to understand what you are saying and to figure out whether there can exist such a thing as a general financial model.'

Fortescue was delighted to hear these words. He gladly lent Schott his handwritten notes with the figures and curves he had shown him. Obviously, thought Fortescue, the mathematician was already half convinced and he was pretty sure the numbers would speak to him.

After a few days trying to fit and extrapolate data in different ways, Schott was very disappointed with the result. The only rule he discovered which seemed to apply fairly consistently to the data the lieutenant had accumulated was a twenty-four hour cycle. In an average market, it seemed to be fairly safe to assume that shares were worth a little more in the evening before close than in the morning, at the opening of trade. Apart from this fairly regular behaviour, the fluctuations he observed seemed completely random, despite what Fortescue claimed to have unearthed. In fact, Schott was reminded of something he had often heard Rutherford say, which seemed to be very true of the data he had before him:

'If your experiment needs statistics, you ought to have done a better experiment.'

So, as agreed, he handed all the data back to Fortescue and apologized for not having had the time to look at it all in detail:

'I have health problems at the moment,' he said, 'and I have already given up most of my mathematical work.'

This was not true, of course, but it was easier than getting into a long and inconclusive argument about trends and projections based on such random data. Nick Fortescue departed a little hurt, but there was not much more he could say.

The incident had entirely slipped from his mind when, about a year later, he opened a newspaper and saw the name Fortescue splashed across a headline. It was an article about a sensational crash in the City. A prominent speculator who had achieved unprecedented success in his investments had suddenly come to grief and lost, not only his own money, but the life savings of many others who trusted him, and even some banks, apparently taken in by promises of huge returns. This adventurer, a military man by the name of Fortescue, had an impressive war record which made him appear to be honest, but had turned out in the end to be nothing more than a con man who stole most of the money he was entrusted with or used it to cover his own debts. To make matters worse, a well-known investor in the City had filed for bankruptcy and, shortly after the Fortescue scandal became public knowledge, committed suicide. Journalists wrote editorials about the story. There was public outrage that a former soldier had exploited his military reputation to gain credit from banks and investors, which was regarded as particularly dishonourable. A man who had fought in the war should have respected his own comrades in arms and not tarnished the reputation of the armed forces by bandying his rank to convince others of his respectability.

George Schott was not too convinced by the sense of moral outrage displayed in the press, but he realized that Nick Fortescue had almost taken him in with his tall story about mathematical modelling. Had he been tempted to construct some kind of algorithm or extrapolation as the lieutenant had suggested, he might even have ended up as an accomplice to some form of embezzlement. With his German-sounding name thrown to the press, that would have been a pretty unpleasant experience.

Luckily, his instinctive caution and a certain degree of humility had saved him. He had not felt able to find general laws for markets and had not quite believed in Fortescue's curves. This did not imply a condemnation on his part, but he did not want to play such games. As he finished reading the newspaper article with a short biography of Fortescue, he was amused to remember that 'old Nick' refers to Machiavelli and is still used by some as the name of the Devil.

It was, in the end, quite consistent that the young soldier used his record to embezzle others. The devilry involved in financial speculation and the devilry of warfare were, after all, not so far removed from each other.

14

Disappointments
after Peace

During the Great War, nothing much happened to further scientific research. It was a fallow period. However, there were advances in understanding. Some scientific papers, which passed almost unnoticed before the war, gained an awful lot in importance through a mysterious process of re-assessment, due perhaps to a change of epoch which modified scientific opinion and eradicated opposition to a number of new ideas. The readers and the researchers obviously had the time to reconsider the literature and had revised some hasty judgments. A new order was beginning to emerge.

Thus, the mythical series of papers published by Albert Einstein in 1905, which had found few readers when they originally appeared, were now considered supreme achievements of human thought. The theory of relativity, which had once seemed so obscure, was saluted as Einstein's crowning achievement. It was hailed even by the general public as the greatest advance in physics of the century. The discovery was regarded as essential even by those who did not quite understand what it was about or what its implications might be.

The only papers which did not benefit from these reassessments were those of George Schott. Somehow, he had not collected many readers even during the Great War and his book passed by unnoticed, despite all the efforts of the lobby in Cambridge which supported its publication. The reason was very simple. General opinion was now in favour of the Planetary Atom, sometimes termed 'the Rutherford–Bohr' model or, more frequently, 'the Bohr Atom', which now appeared as a decisive step towards a new and very advanced form of mechanics, still to be perfected. There were mutterings about an undulatory theory, while some preferred to use the word 'quantum'. Whichever was more correct, Bohr had taken the plunge and was clearly the initiator of a new physics, so he gained enormously in prestige.

As a result, Schott became progressively more isolated. This did not really discourage him, as he had always worked alone ever since the mental decline of his brother. He knew that fashions come and go and that opinions often change. What was trivial before can suddenly appear important and the celebrities of today are often gone tomorrow. The vulgar and uncertain trumpets of fame were not for him, and he had decided a long time ago not to take any notice of them. He trusted in a more distant future and spent his time perfecting the details of his theory, making it even smoother and more elegant in form and adhering ever more closely to the conceptual framework elaborated by Maxwell.

He was confident that, by following this narrow path, he would arrive at a formulation so obvious and so profound that nobody would be able to resist its power. However, events just did not seem to conform to his master plan. Somehow, fate seemed to have decided otherwise.

A long string of eminent researchers, most of them German, Planck, Kramers, Krönig, Ehrenfest, Jordan, Heisenberg, Born, Schrödinger and others were defining the spirit of the times and doing so in a way which seemed to support Bohr's way of thinking. In fact, Bohr was, in his turn, becoming the pope of this new religion after Rutherford, despite the vagueness of some of his statements. Even the serious and solid physicist Unsöld, whose work was more congenial to Schott because he had not fallen for newfangled ideas, was gradually won over to the quantum theory, or at least to its language.

Only Schott's hero, Albert Einstein, continued to express reservations, despite his strange and rather illogical idea of introducing photons as the quanta of light. Nonetheless, for the lone sage of Aberystwyth, whose daily dialogues were with the waves of the sea, Einstein alone still spoke with a voice of reason.

George Schott, against his own will, had been propelled into the position of the leading British opponent of the new quantum mechanics. This was not at all what he wanted and rear-guard actions were not his cup of tea. He was expected to join the fray and indulge in antagonisms quite foreign to his character.

He tried to understand why he was being pushed into this corner and even wrote a letter to Louis de Broglie, because he knew that the French physicist, like him, was a great admirer of Albert Einstein and had expressed doubts of his own about this pervasive new theory.

De Broglie, who always answered letters very politely and in great detail, sent him a reply a few weeks later which covered many of the points George Schott had been worried about. In this reply, de Broglie expressed his own anxiety, as he felt close to being submerged by a huge wave of incomplete scientific papers containing highly debatable assertions and many half-truths. He wrote:

At the Solvay Conference in 1927, I attempted in my own way to resolve the problems you raise. However, my own presentation did not have much impact. Pauli made many objections. Schrödinger did not believe in my corpuscles. Born, Heisenberg, Pauli and Dirac were developing their probabilistic inter-pretation which is now becoming the orthodoxy. Lorentz and Einstein wanted to save classical physics. The only one to give me any encouragement was Einstein, but even he did not express firm approval for what I was doing…

As for Bohr (and you remind me of my own words about him), he is indeed the Rembrandt of contemporary physics, as he practises a kind of 'chiaroscuro'. He wanted to redefine my waves. They would then lose all physical meaning and would merely represent some kind of probability flow. This would be personal and subjective and would be instantly altered whenever the user receives new information. The real question we are ultimately left with is the one stressed by Einstein. Either the so-called 'orthodox' interpretation is a complete description of reality, in which case determinism would no longer apply on the atomic scale, or it conceals within it (as do the old statistical theories of classical physics) a perfectly deterministic reality of which only the full set of parameters is inaccessible to us.

Einstein, to whom Schott also wrote, responded very much along similar lines, while also advising Schott against any attempt to stem the tide. In his view, opposition to the new fashion was useless, since times had changed and mathematical rigour was no longer the name of the game.

He also added in his letter a comment in the style of an aphorism which Schott found quite incomprehensible:

As far as the laws of mathematics refer to reality, they are not certain; and as far as they are certain, they do not refer to reality.

This was a very odd remark. Schott could make neither head nor tail of it. In his own opinion, mathematics and reality were but one and the distinction between the two seemed in itself paradoxical. He concluded that the great Einstein had penned this somewhat incoherent statement in an attempt to cheer him up and just took it as a sign of friendship.

Ehrenfest, whose support Schott also sought, responded simply with a statement that quantum mechanics had no real future. It was a temporary aberration from which physicists would recover. It was better to let it pass.

Trying to prevent it, he added, would make no more sense than attempting to prevent the birth of a new religion.

All of this was very different from the somewhat idealistic vision of science George Schott had harboured since setting out to study theoretical physics. The war had obviously changed people, and not necessarily for the better. He could hardly recognize the world of research as he had known it. Fortunately, there were still a few fixed points one could hang on to, and one of these was of course Maxwell's theory of electromagnetism. So, Schott hung onto that and decided to let the rest go by.

Rutherford left Manchester and reestablished his research at the Cavendish Laboratory in Cambridge. The 'boss' was broad-minded and had no difficulty welcoming Schott as an old friend whenever he chose to pay him a visit. This was despite the unenviable reputation Schott had been foisted with of being the main opponent of the new quantum theory.

At this point, Ernest Rutherford was pretty well convinced he and his friends had won the argument. He was taking a more relaxed view of science in general and even accepted the opinion that theory might be judged on its aesthetic merits to some degree. Indeed, he made a somewhat surprising and almost lyrical defence of the beauties of mathematical physics before the Academy of Arts in London:

'I think a strong claim can be made,' he said, 'that the process of scientific discovery may be regarded as a form of art. This is best seen in the theoretical aspects of physical science. The mathematical theorist builds upon certain assumptions and according to well-understood logical rules, step by step, a stately edifice, while his imaginative power brings out clearly the hidden relations between its parts. A well-constructed theory is in some respects undoubtedly an artistic production. A fine example is the famous kinetic theory of Maxwell...the theory of relativity by Einstein, quite apart from any question of its validity, cannot but be regarded as a magnificent work of art.'

This was a very new and unusual point of view coming from Rutherford. Of course, he did not really broach the question of what should be done about a mathematical theory which might be perfectly beautiful but nonetheless, in his opinion, incorrect. What should one do with a work of this kind? Was it to be included in the great history of science, cast aside as some kind of waste product or hung up on the wall of some museum? One can assume that Rutherford's little speech was intended to reassure the artists residing in the place in which he made the address. However, Schott could still take some comfort from these words. Rutherford had evolved in his attitude towards mathematical beauty. Perhaps he might even consider that the treatise on circulating electrons had been of some value in the development of the subject.

Many events had taken place in the intervening years. Ralph Fowler, seriously wounded in battle, had returned in a sorry state. Despite his condition, he took charge of the theoretical physics group in Cambridge. Rutherford had tried desperately to attract Niels Bohr, and had formally offered the position to the Danish scientist, who enjoyed a worldwide reputation. Alas for the United Kingdom, this was not to be. For Denmark also, holding on to Bohr was a national priority. It was a question of pride in its own culture to retain him at pretty well any cost. He was made such a generous offer that he turned down the position Rutherford had obtained for him in Cambridge and remained in Copenhagen, much to the disappointment of 'the boss', who was forced to offer Fowler the job.

As the crowning achievement of his brilliant scientific career, Ernest Rutherford was now in full control of a formidable scientific empire: the Cavendish Laboratory in Cambridge. These were to be the golden years of British physics. Nothing of the same importance had happened, it was often said, in the same place, since the work of Isaac Newton. Twelve of the former students and collaborators of Rutherford were to be awarded Nobel prizes in close succession. It almost seemed as though, to be even considered for such an international accolade by the Swedish Academy of Sciences, it had become a prerequisite to have worked with Rutherford.

That was, unfortunately, not enough if you happened to be called George Adolphus Schott. Somehow, he always seemed to be omitted, or out on a limb of whatever was considered important in physics. It was true that he had decided long ago to take no notice of international glory or even of national reputation. Nonetheless, he could not help feeling a little frustrated at being so close to something he could never actually reach. For this reason, he valued his personal connection to Rutherford and was happy at least that the friendship they had developed remained true in spite of everything which might have separated them.

'You see,' he said to Eileen when he saw her again, 'I am back and I am so happy to be with you after all these traumatic years of the war. Believe it or not, I even became a captain in the British Army, but I can't imagine why, as I never did any fighting. And your father is a wonderfully generous man who never let our little scientific tussles intrude in our lasting relationship. We have agreed that we can disagree and remain, as we always were, the best of friends.'

As it happened, Ernest Rutherford was just behind them and caught the words of George Scott. He responded immediately in his great booming voice:

'My daughter admires you very much,' he said, 'and she is completely right to do so. You are one of the few people in this world who are faithful to their intellectual principles and hold fast. We are not always on the same side of the barricade, but we are honest fighters and respect each other for that. When we disagree, I admire you for your loyalty and for your sincerity. And I also salute you as the true British gentleman you are. That is the highest of all qualities in my eyes.'

For Schott, these words of recognition had more value than any prizes or public distinctions. He was so touched that he could not even find words to express his gratitude.

Later on the same evening, he was able to ask Eileen what she thought of all the great changes which had happened during the war years. Then, without any direct reference to what her father had said, he added:

'I suppose your own opinion may not be quite the same as your father's. What do you think of all these new judgments and reputations, which seem so different today from what they were before?'

'That is not for me to say,' she replied, 'my opinion matters very little. My father says there are two kinds of heroes now. Those who went into battle, saw action and have returned, to whom honour and glory are due, and those whose battles were for intellectual victory, in the interest of future generations. He says they should not be forgotten either.'

'But what is your own opinion about that?' asked Schott, who could not help wondering about Eileen's feelings. 'Who are the real heroes at the end of the day?'

'I feel the same way as everybody does, Adolphus,' she replied in a soft voice. 'These two forms of heroism should not even be compared, despite what my father says. Those who return from the war obviously have priority in our minds. Some of them will not survive for long and we must preserve their memories. Otherwise, they are the ones who will be forgotten. Their names, and those engraved on monuments must obviously be remembered. The others have a huge advantage. Their names will survive even without any help from us. Sometimes, the names of great scientists even become part of our language, like Watts and Volts which are in daily use. That is another kind of fame.'

A few months after this conversation, Eileen announced her engagement to Ralph Fowler, who was to become, through their marriage, a member of the Rutherford family. She was prepared, like many women of her generation, to lead the difficult life of a wife who is also a nurse. Schott was much aggrieved by this unexpected news but, of course, was fully aware that something like this had to happen sooner or later given the age difference between them. This was the only real complaint he had against his own destiny. He had been born too early. He tried to remember all the sayings of wise philosophers about fate and how useless it is to rebel against ill-fortune.

The gods were perhaps too generous in their gifts, he reasoned. *They brought me the abilities I have, the means to develop my small talents and subjects to work on, which I can cherish. The only thing they got wrong was my date of birth. They chose the wrong day of the wrong year for me and that was unkind. I should have been born twenty years later than I was.*

Whenever he admired nature, Schott spoke of God in the singular, to attribute all the marvels to a well-intentioned supreme being, but when he was critical of destiny, he preferred to refer to the gods in the plural. Pagan divinities were less worthy of respect and he felt it somehow less of a blasphemy to speak that way.

To overcome his disappointment, he decided to work on the most challenging problem he could think of. He would find the symmetrical distribution of charges and currents Ehrenfest had hinted at, which can rotate around a fixed point without violating the principles of Maxwell's theory and without radiating, despite its circular motion.

If I find it, he told himself, *I will have achieved a classical distribution of charges which can behave in the same way as Rutherford's electrons, but without violating any of Maxwell's theorems. That will be the ultimate solution, the one everybody is waiting for to rehabilitate classical physics.*

He felt the need for a strong challenge, a bit like the famous theorem of geometry the French mathematician Blaise Pascal always kept to hand to deal with toothache. He needed something really difficult to occupy his mind and fend off feelings of pain. He tended to put it down when he felt a bit better, much as one should stop taking opium when it is no longer needed. That way, the mathematical treatment lasted longer. His aim was to keep it, if necessary, for a few years, because it was the only medicine he knew which would cure his particular ailment and he needed no doctor to be confident of that. So, it was best to use it only when absolutely necessary.

He kept a few pages with some of the key equations on a writing table next to his bed. It was in fact a night table, not intended for writing at all, but he sometimes found it useful when he woke up in the middle of the night with a good idea in mind.

Schott was never in any hurry — as Rutherford would have been and remained all his life — to publish his results. Schott was not at all worried about competition. Indeed, there was none in his area of business. He could afford to chew his pencil as long as he wanted, in the full knowledge that he was probably the only person in the world working on the problem at that particular moment in time.

The question of non-radiating rotating charges, which Schott referred to occasionally as his painkiller, was constantly to hand by his side. He always kept the most recent development with him, so as to be able to take it out and read it whenever necessary. On one occasion, as he was strolling down the seafront, he nearly lost part of his work when a gust of wind suddenly wrested a page from between his fingers and tore off several equations.

15

His Cousin Settles in Aberystwyth

George Schott's cousin from Frankfurt, Charles Jakob Schott, who now wrote his name as Jacob, came to visit him in Aberystwyth. He had never been to the town, but Adolphus always praised it, and eventually his cousin felt he needed to discover what it was like. As he was keen to form his own opinion independently, he asked George not to meet him in London as he had offered to do, but to let him travel alone all the way to Wales.

On his arrival, Charles Jacob was indeed delighted by the climate along the seaside, the clean air and the prevailing westerly wind which seemed to keep all of the British industrial pollution at bay. It was indeed a most charming place, quite unspoiled, and even the seaside houses, despite being of recent construction, were pleasing to the eye.

'Landscape painters should travel here,' he declared. 'This is much better than where the rest of your family is living.'

In fact, he was so pleased with Aberystwyth that he searched for a few pictures he might take back with him to hang on his wall in Frankfurt as a souvenir of his trip and remind himself of where his illustrious cousin, George Adolphus the Professor, was living.

They were accustomed, every time they met, to begin with a game of chess. George Adolphus, of course, was a famous player. Actually, the two cousins had learned the game together and taught each other many subtle moves. Charles Jacob, in Germany, had not found such good opportunities to play at a high level as his cousin and did not belong to any recognized team. He did not enter for official competitions, but he remained an excellent player, almost as good as his cousin. Also, as they had played each other very often, they knew each other's psychology and recognized each other through their favourite moves.

On this occasion, however, Charles Jacob had no desire to play. He seemed much too much preoccupied by events.

'I don't feel up to a game,' he declared. 'I must talk to you about the situation in Germany. It is truly catastrophic.'

And he began a long soliloquy about a whole range of problems George Adolphus had never heard of. This was not the Charles Jacob he had last met. Something dramatic must have happened to change him so. Presumably, what had occurred was what he was complaining about, but all of this was completely new to George Adolphus. These problems were quite unfamiliar. He would never have thought the situation was so different in Germany. For him, the war had been an episode in the past. When it was over, he had returned to a somewhat different professional life, but the rest of society had resumed its previous ways and he had not noticed any fundamental change.

'You can't imagine how lucky you are,' said Charles Jacob by way of introduction. 'You are living safely, miles away from the awful situation we are now facing in Germany.'

'Your happy island of Great Britain is a rich and peaceful land. You have none of the problems we must contend with. To give you some idea, let me just tell you about the price of bread and how it has multiplied in recent weeks. The money we earn seems to be worth nothing at all by the time we reach the shops. In the time it takes to walk to a bakery, the price of a loaf has increased and if you have at all to queue, it is no longer the same by the time you get to the counter. Money is no longer worth anything. You must spend it as fast as you can. If you wait, you are soon ruined. And don't let me tell you what people are beginning to talk about. We are within a hair of a huge revolution, something really cataclysmic. Before long, many people will have lost everything and, when that happens, I think there will be blood on the streets.'

Charles Jacob talked in this way for some hours and, for fear he might seem to be expressing unbalanced opinions, gave Adolphus a short book to read, by the great Austrian novelist Joseph Roth. It was *The Spider's Web*.

'I know you don't read novels,' he said, 'but this one is just different. If you want to know what is really happening in our homeland, you can't avoid reading this book. It will tell you all, not only about our present plight, but also about our likely future.'

Schott never read literary works. He was just not attracted by fiction. But he respected his cousin's opinion and, of course, was curious to discover what was really going on in Germany. So, he agreed to take it and to start reading that very evening. For the first time in his life, he experienced not to be able to put a book down and was still turning the pages over in the early hours of the next morning. He had discovered an unknown land, the place from which his parents originated. It had always been painted as a nation of culture, sometimes misguided, but never barbaric. He was discovering a new reality, something he had never suspected before. His Germany had been quite different. He began to understand more clearly what had been going wrong in Europe and some of the horrors his parents had probably witnessed at first hand, although they had never wanted to speak about them.

'How can this all be true?' he asked his cousin, 'and how can you manage to live out there with all this going on around you?'

They began a long conversation which lasted through the following day. George Schott learned from his cousin about German politics. He discovered the most unpleasant facets of current affairs, aspects of power struggles he had never really concerned himself about. In the depths of Wales, why should he have bothered about them? When it had been required, he had done his duty and joined the war effort, but had never really asked himself why he was wearing a uniform. It had just been what his country asked of him. He fully understood that by being a British subject he had duties as well as privileges. Here, however, was something quite different, something really disturbing. It was a new reality he had not been conscious of. He began to realize why a war had been fought. In fact, he also suspected there might even be some need in the future to start the whole business over again, though what the consequences of that might be, he shuddered to think.

When his cousin left him in the early hours of the morning to get some sleep, George Schott retired to his bedroom. Quite by chance, he saw some of his own visiting cards on the desk. He had always had a problem with his second name 'Adolphus'. And there it was again, written plainly on the cards for all to see. Why had he kept it? Until then, he had only found it slightly quaint and out of place, but he had accepted it as a part of his personality — something left of his German heritage. Now, he was suddenly horrified by the sight of the name. There was no way he could continue with it. He reached for a pen and carefully blotted it out from all the cards he found lying around. He did not want to share this name with someone whose existence he had just discovered.

He would have to get some new cards printed right away. But what should he do about his second initial? After some hesitation, he remembered the name of an old uncle in Rheinland, now deceased. Yes, that would do the trick. Henceforth, he would become Augustus. George Augustus Schott. That sounded similar, but was entirely different.

I will have them printed tomorrow, he said to himself. *Most people won't even notice the change, but I will feel much better about it. In fact, if anybody does notice the alteration, it will make them think, and wise people will understand what it means. I can live with that. I don't think I need to give any explanation.*

His cousin Charles Jacob thoroughly enjoyed the stay and showed signs of becoming very attached to Wales. He commented that the quality of life there was remarkable. He had also much appreciated meeting a young lady from the town, whom he had befriended while walking in the countryside near Aberystwyth. This, no doubt, was the reason for some further visits. He fell into the habit of returning fairly often, ostensibly to see his cousin, but also for more personal reasons, despite the long and rather tiring journey by train and boat all the way from Frankfurt.

A few years later, he married the young lady and settled in Britain. He no longer felt at home in Germany and this had been apparent for some time. He explained that the rising xenophobia in his country had driven him away.

'Did you know,' Charles Jacob said to his cousin, 'that the physicist Philipp von Lenard whose work you admire (and who, by the way, is not truly German but Hungarian) has joined that group of horrible German nationalists Joseph Roth was writing about? Even the intellectuals in our land now have indecent opinions. Lenard says that Albert Einstein's work is "Jewish Physics". Can you imagine anything so stupid? Maybe he thinks that the 'von' in his name is not enough to convince the public he is a real German. That is probably what drives him to such extremes.'

Schott was horrified by this unexpected news. It was true. He had admired many of Lenard's papers and was unable to understand how an important scientist, a Nobel laureate of high reputation, could sink so low. Although he was born in England, George Schott had sometimes felt a foreigner because his parents came from elsewhere. Now, he realized his life would have been much worse if they had decided not to emigrate from their homeland. Why they had left had never been completely clear to him, but he had discovered thanks to his cousin how much they were right to go. How strange that Germany should be the birthplace of so many great physicists he admired and, at the same time, a country so devoid of elementary common sense!

George Schott, who had always found the right balance between the two cultures he knew so well, found it very hard to adjust to what his cousin had just taught him. It was as though he had been standing on stilts and one of them had just been kicked away. Again, it was mathematics that came to the rescue. At least symbols were impartial. It mattered not too much in what country they had been invented. Once they had been accepted into the great family of scientific usage, they no longer belonged to any particular culture. They were simply international and above all the meanness of men. Science was a better world. Maybe that was what Rutherford had meant when he said that splitting the atomic nucleus apart was in reality a more important event than the Great War.

Against all odds, Schott was still working away discretely at the task he had set himself. Meanwhile, Eileen's husband, Ralph Fowler, had a brilliant career in Cambridge as the professor of theoretical physics. He had many famous students. One of them was Paul Dirac, who almost managed to reconcile the principles of quantum mechanics with Einstein's Special Relativity. Schott followed his attempt with great curiosity. This at least was a noble task and he was not surprised when Dirac was awarded a Nobel Prize, although he remained essentially unconvinced that the two theories had been really harmonized. Dirac had also said something very dear to the heart of Schott when he reminded everybody that equations must be beautiful or else can surely not be true.

The aesthetic side of mathematical physics had always been one of Schott's preoccupations.

No doubt it was in spite of what she had always heard from her father against theorists and the unreliable stories they tell that, in 1921, at the age of twenty, Eileen married the theoretical physicist. Ralph Fowler, like Rutherford and Schott, was also from Trinity. She had four children by him but died very young, in 1930, nine days after the birth of the last one. Mary Rutherford remembered the fondness Eileen had as a child for George Schott and sent him a printed announcement with the sad news.

Schott was affected well beyond any feeling he might have been able to express. This was something he never realized might happen. In his mind, she was so much younger than him, he was sure to depart this world before her. He sat on a bench by the seaside and tried to come to terms with the sudden news. For the first time in his life, he felt there was something completely useless about everything he had attempted to achieve. Maybe even higher mathematics, after all, was not very important. The vanity of all things somehow became clear to him, although he had never really thought about this idea before.

He walked a little further along the seafront and sat on another bench with the letter he had just received in his hand. It was hard to believe. Then, he remembered that he had kept a carefully folded letter in his wallet. It had remained there all these years, simply because there was no better place to keep it. It was the letter Eileen had written to him so long ago.

He had never been able to answer it. He saw that now quite clearly. It was in some ways a painful memory but, on the other hand, if he had been capable of finding the right words, what would it have led to? Even if it had been possible for him to answer the letter instead of leaving what she had written up in the air and never really alluding to it when they met again, nothing could have been resolved. It would just have made matters worse and rendered any further relationship between them quite impossible to continue.

So, after all, it was a blessing that nothing had happened.

He took out Eileen's letter and was surprised to discover that it was exactly the same size as the announcement card he had received in the morning. So, he slipped them both together inside the same envelope. There was a small flat stone, round and very white, lying on the ground next to him. It seemed to be waiting. He picked it up and slipped it in as well.

Then, as a child would have done, he threw the envelope out to sea with all its contents and watched for a moment as it disappeared into the spray of a large wave. He stayed there for a while wondering what he had done and why he had done it. The conclusion he reached was that, after all, his life only made some little sense through mathematics.

As for Fowler, whom Rutherford had only taken on as a second choice into his group when Bohr turned his offer down, he was now becoming one of the great stars of physics in Cambridge. George Schott followed all the details of his impressive career in scientific literature and in the press. In 1922, Schott had also been elected Fellow of the Royal Society but now regarded himself as an unimportant professor in the small provincial town of Aberystwyth. So, he felt somewhat out of place in such an august body and never attended meetings. Quite apart from anything else, travelling to London had become very tiring. Shortly after Eileen's death, Ernest Rutherford became a Lord and George Schott wrote him a personal letter of congratulation.

In 1933, George Schott published his last paper, in which he described a spherical configuration of electrical charges and currents with remarkable and unexpected properties. It is completely consistent with Maxwell's equations but, nonetheless, able to rotate about a centre without radiating. Such behaviour had seemed contrary to the principles of classical physics until this configuration was discovered. It took Schott many years to find it. The quest had become almost an obsession.

Unfortunately, however, this configuration was of no particular use as it did not describe the properties of any real atomic shell. It was in itself an abstract and marvellous construction, a *tour de force,* but posterity, in the end, just kept it for amusement along with other mathematical curios, locked away in one of those curiosity cabinets which used to be fashionable, but were very rarely opened once the key had been put away.

16

Schott Passes Away
and a British Bomber
is Shot Down

Despite the ocean's healthy breeze bearing invigorating iodine and pure oxygen from the wide open spaces where nature had not yet encountered anything like pollution, George Adolphus Schott, as he became older, began to feel something slightly painful gnawing away at his lungs. It was not new. He had already experienced this unpleasant feeling before and perhaps he even remembered it from his childhood in Bradford. Sometimes, for example when he travelled by train and the wind blew heavy smoke from the locomotive into the carriages, the familiar smell of soot and sulfur triggered a persistent cough. And sometimes, even without any smoke, just thinking about it as he walked along the seafront was enough for the cough to reappear. In fact, he never seemed able to be rid of it. It was there now, all the time.

Coughing also brought back memories of the polluted environment of the family factory in Bradford. The worst experiences in recent times had been when travelling between London and Manchester. Luckily, he had little reason to catch trains now. He no longer went to the capital any more, but the memory of smoke-filled carriages lingered on, especially the smell when the train suddenly entered a long tunnel and the air pressure changed.

To stay in touch with friends, Schott took to writing letters, something he had never been very good at but which had now become necessary to him. Posting a letter was a lot easier than trying to catch a train. So, he taught himself to include the kind of unnecessary personal details people like to find in letters. He had noticed it was more likely they would be read and might even elicit answers. He still remained rather clumsy in what he wrote, but his friends knew him well enough to make allowances for his peculiar style. One who usually answered in detail and kept him informed of events was Mary Rutherford. Maybe he reminded her of happier times when her young daughter had confided her fondness for the strangely reserved mathematician.

George Schott now needed a walking stick for his daily constitutional along the seafront. He liked to walk with his eyes fixed on clipper ships carrying precious cargo towards the Americas or beyond. Amongst the elegant sailing boats, which still carried freight in those days, were more and more black vessels like clumsy crows with tall chimneys spitting out smoke which seemed to be intent on creating new clouds in the sky. He knew very well what kind of smells could spill from such a chimney but, fortunately, the boats were far away and the wind carried evil odours high up into the sky so that none actually reached him. A benefit of living so close to the open sea.

Fortunately also, such boats were still in the minority and the ocean ahead provided enough wide open space for them to travel further towards the horizon, intent on spoiling the red skies of sunset with their ugly billowing black and red froth.

Soon, fresh air and a sea breeze no longer sufficed to serve as an antidote to George Schott's persistent respiratory problems. The cool and humid autumn wind, from which nothing seems able to protect ears and lungs, began to penetrate further, cutting into his throat and provoking more and more violent spasms of coughing.

Instead of fighting the weather and walking briskly against the wind, he had fallen into the rather bad habit of sitting quietly on a bench facing the open sea and meditating on the seafront without realizing how cold his body was getting. His mind was full of beautifully symmetric mathematical objects which he alone could see. There was really nobody around him to whom he might have explained just how magnificent they were. They unfolded before him like complex flowers and he enjoyed rotating them in his mind's eye. Sometimes, when he felt he had reached a significant result, he would take a folded piece of paper out of his pocket and hastily jot down essential points.

Little by little, George Adolphus Schott was losing touch with reality. He no longer took letters all the way to the Post Office because most of his correspondents were lost to this world or no longer able to write.

His new dialogue was with a very old copy of Maxwell's *Treatise on Electricity and Magnetism*. The cover had dropped off a long time ago and this indeed was rather convenient, because he could slip it into his coat pocket before going out for a walk. So he kept it that way. A few outer pages were missing too, but they were not the important ones. Others, earmarked several times over, had turned quite yellow with age, as old paper does, and were foxed by the damp weather. They were so covered in marginal notes that there was no space left to write anything more. So, with his chewed pencil, he was obliged to insert further comments in between the printed lines. They were only just visible.

He was holding this old book in one hand when he fell off his favourite bench one winter morning in the year 1937. Probably, his mind had been struck by an inspiration he had no time to translate into symbols and reduce to an equation. It must have been still more abstract than usual. He was 69 years old. Passers-by tried to assist him and paid no attention to a few sheets of folded paper he had dropped. The wind carried them away in a spiral motion towards the sea.

Perhaps, if one of them had been saved, it might have revealed some final secret about rotating charges which manage not to radiate energy as they turn. However, since none of the people around him understood anything about higher mathematics, it is more likely that it would have ended up, as so much written paper does, directly in the dustbin. That is what usually happens to scribbles people don't understand.

Only a young boy, the son of a bartender who worked near the seafront and obstinately kept his establishment open even out of season, remembered the old man who had often kept him company when his father's premises were empty. Schott had taught him a complicated game with strange rules which stuck in his mind. He continued to play it throughout his life and was grateful to the stranger who had taught him the rudiments of chess.

Schott passed away in 1937, exactly the same year as Rutherford, after quite a distinguished career as a professor in a tiny provincial university. He left £6,427 17s. and 10d. in his will — a reasonable sum, but not very much.

It is quite possible that his reputation as a chess player was greater in his lifetime than as a scientist, even amongst his physicist colleagues, although astronomers remembered him better for some equations.

Very few people even noticed the passing of George Adolphus Schott. Compared to the (almost simultaneous) disappearance of Lord Rutherford, it was indeed a very low key affair. By then, it was considered that Schott's objections to the Planetary Atom, for some reason, were simply ill-founded. The peculiar twist to his story was that his results, in fact, were not wrong at all. He had simply worked them out under inauspicious conditions.

To make a mark in science, several conditions must be satisfied. First — as Albert Einstein himself pointed out:

'One needs an idea which seems shocking at first, gives rise to controversy, and eventually turns out to be true.'

Schott's situation was the exact opposite. Everyone had objected, both to Rutherford and to other would-be reformers of classical physics, that electromagnetic waves, by radiating energy away, made their models unstable. Schott had merely confirmed their conventional wisdom. So, that was already a poor start. His results were in full agreement with the generally accepted view, and there was no surprise. The Planetary Atom of Rutherford, on the other hand, was initially quite shocking and it had taken Bohr's nerves of steel to establish the new theory bearing his name. So, on the basis of this first criterion, Bohr and Rutherford were clearly well-placed to become famous, while Schott was destined to be forgotten.

The second condition to satisfy is less obvious but equally important. The new idea must appear at precisely the right moment in history. Neither too early, nor too late. This is where men's reputations lie in the lap of the gods. Inspiration must come at the right time. There again, Schott was badly placed.

Had he thought of his problem ahead of the Rutherford atom, his work would be in all the textbooks as a classic result obtained along the way towards the new mechanics. He would have been one of the leaders. However, because he came after the hypothesis of the Planetary Atom and argued against it, he found himself inevitably in conflict with Rutherford, whose latent hostility was a big problem. Despite his sympathy for Schott, Rutherford could not help feeling that the treatise on the radiation of circulating electrons had stopped his progress and thwarted his ambition to create the basic model of atomic structure.

The huge and innovative contribution of George Schott was destined to remain in limbo for yet a third reason, still outside his control. He had formulated his theory too early. This is perhaps the strangest part of the whole story.

He had discovered a property which is, today, called 'synchrotron radiation'. However, finding it when he did was a bad move. This is the radiation emitted by charged particles accelerated on circular orbits in laboratory instruments termed accelerators, for example, electron synchrotrons or storage rings. The accelerated particles are regarded as 'free' in the sense that they are not subjected to the constraints of quantum mechanics. Consequently, they do indeed radiate away their energy, exactly as predicted by George Schott.

However, there were no accelerators at all in his time.

Astronomers did detect radiation due to the circular motion of electrons, but it was very far from our planet, very distant from any laboratory, rather weak and quite useless. It was far from interesting compared to the great technical developments made during the Second World War. Schott had simply picked the wrong moment to make his great discovery. His timing could not have been worse. Not only had accelerators not yet been invented, but two world wars diverted attention away from pure science.

On the 19th of May 1944, a Lancaster bomber loaded with eleven 500 kg bombs and three 250 kg bombs took off from the military airport in Waddington. Its mission was to destroy a railway junction at a place called Revigny-sur-Ornain in the Meuse area of France. The pilot was Sergeant Keith Jacob Schott, son of Charles Jacob Schott and Bertha

Branee Schott, born on the 18th of January 1924, who had enlisted with the Royal Air Force at the age of eighteen. During its mission, the bomber was attacked by a German fighter plane and shot down in a field, close to the village of Brabant-le-Roi. There were no survivors.

The inhabitants of a nearby farm recovered the bodies and gave them a respectful burial as befits soldiers killed in action. Later, the name of Keith Jacob Schott was included in a list of all the Jewish soldiers of the Royal Air Force who died in the Second World War. His entry reads as follows:

Born 18/1/1924. Son of Charles Jacob SCHOTT (1870–1929) & Bertha Branee née SAGEL (1885–1957), of 12 Avoca Street, Elwood.

Comment: Airborne 23:01 on 18th of July 1944 from Waddington to bomb a railway junction at Revigny. Bomb load 11 × 1000lb, 3 × 500lb bombs. Shot down on its bombing run by a night-fighter, crashing in a field near Brabant-le-Roi (Meuse), 2 km north of Revigny-sur-Ornain. Those killed are buried in Brabant-le-Roi Churchyard.

Surname: Schott First Names: Keith Jacob Rank: Flight Sergeant Service No: 419594 Date of Death: 19/07/1944 Hebrew Date: 28 Tamuz 5704 Age at Death: 20

How Died: Flying battle. Lancaster PB234/Lancaster

R5485 was lost in the same raid

Where Died: North West Europe Cemetery: Brabant-Le-Roi Churchyard, France Service Details: RAAF attached to 467 Squadron, RAF

Served Occupation: Clerk

Age at Enlistment: 18

Chapter

17

Chapter

Patrick Blackett
Never Mentioned
Schott

After the Second World War, Professor Patrick Blackett (often referred to as 'PMS' by his initials), Nobel Laureate, personal friend of Churchill and illustrious naval officer, took charge of the Physics Department at Imperial College with the firm intention of raising its reputation to that of a premier institution not only in the United Kingdom, but worldwide. He had become one of the most influential scientists from the famous team who had worked with Ernest Rutherford. He was accustomed to passing on its folklore by citing words of 'the boss' and telling anecdotes from the times, both in Manchester and in Cambridge, where nuclear physics as a discipline had been invented. All the physics students at Imperial knew Blackett's anecdotes by heart.

Even as an old man, he demonstrated the operation of the Wilson Cloud Chamber in the teaching laboratory and it was well-known that, in his absence, nobody else seemed able to make this peculiar and capricious device function.

In the worked examples which were set in his course, there was an important problem which students worried about every year. It referred to the deviation of the trajectory of a shell fired at the Battle of Jutland owing to the Coriolis force. Students learned the solution of this conundrum and analyzed it carefully. It was always likely that it might turn into an examination question, although in fact this never happened. It was also a widespread belief that one should study in all its details a particular nuclear reaction, the first one ever produced by man, an achievement of which Patrick Blackett was especially proud.

The sailor-physicist had become a walking memory of the legendary Rutherford years. As time passed, however, his lectures were becoming less coherent. This did not diminish his prestige, but meant that students pressed into the lecture theatre to hear him for a different reason. It was not so much to follow the course, but rather to pick up snippets and tales about the golden years of British physics. Many of the students (including those from around the world) would say to each other:

'Let's go and listen to the old gentleman. He is completely out of fashion, wears very odd clothes and must definitely be the last representative of a disappearing old English style, but that is exactly why we must go and listen to him. When he is gone, who will be able to take his place? There will just be a gaping hole in the memories of all those who were lucky enough to be taught by him.'

So, Blackett remained a star even a long time after his lectures had lost all pedagogical value and their physical content no longer impressed the students as much, perhaps, as they should have.

He did, however, possess some quite remarkable scientific intuitions well after the Second World War and those who did not take him seriously were proved quite wrong. For example, it was his idea to study the direction of magnetization of rocks in Canada, which seemed a rather crazy activity when he started it. Aeroplanes were loaded with magnetometers and, just to please Blackett, former wartime pilots spent hours of their time flying back and forth over huge expanses of the frozen continent. That is how he obtained the first hard evidence in favour of the theory of continental drift, which until then had been regarded as an idea impossible to verify. He had understood that rocks, when they solidify, trap some vital information frozen in their structure: the original direction of the magnetic pole. So, as the continents drift, the rocks preserve essential data from which one can reconstruct how much they have moved from their initial position.

Immediately after this spectacular success, Blackett decided to close down the rock magnetism research group, much to the dismay of the researchers and their staff. There was some dissatisfaction with what was considered to be an autocratic and arbitrary decision.

He explained why, and his explanation was interesting:

'Now is the time to close the group down,' he said. 'We have proved our point and we can never obtain a more important result in the future. Our researchers are at the peak of their reputations and I have no worries about their careers. They will find excellent jobs in other universities who will be proud to have recruited them. On the other hand, if we kept the group alive, it can only decline and carry them down with it. Surely, that is not what we want. So, we must stop the group right away, not tomorrow or after tomorrow, but immediately.'

It turned out to be a wise decision. Everything worked out exactly as he had predicted. The upset he had caused was soon forgotten and the researchers, indeed, found new positions pretty quickly. Blackett had inherited from Rutherford a directness of approach which helped him a lot in the management of university departments. When important decisions were to be made, he was feared but also highly respected.

He also inherited some of Rutherford's prejudices and a rather selective memory of the events he had witnessed in Manchester and in Cambridge. He had seen not only Niels Bohr going by but also George Schott. However, he never pointed out to students that Schott was the one who had challenged Rutherford's Planetary Atom and who had held back its development until Bohr appeared on the scene. He might have explained the role of George Schott to the students and given some idea of the controversy, even if only to say that Schott was wrong and Bohr had been right not to listen to his objections. But he preferred not to talk about Schott at all. His guiding principle was that mistakes should just be forgotten, that the truth, once discovered, need no more be challenged and that one should point out the right path, especially to students, without encouraging them to get side-tracked. So, the very existence of George Schott and the nature of his objections were carefully removed from Blackett's summary of events. It was more important to study experiments than to worry about the nitty-gritty of theory and that, of course, was exactly the teaching inherited from 'the boss'.

Nonetheless, on one occasion, Blackett demonstrated a deep respect for theory and some intuition about its real importance. It was when he decided to bring a young theorist called Abdus Salam from Cambridge to Imperial College. Theorists are necessary in a laboratory, but must be chosen carefully. That was also a lesson he had learned from Ernest Rutherford.

18

Synchrotron
Radiation Came
Too Late

The first cyclotron, an apparatus designed essentially for the purpose of accelerating protons on circular orbits, was constructed by Ernest Lawrence and Stanley Livingston at Berkeley in the United States between 1931 and 1932. At the time, the energies achieved were still rather low, and losses due to radiation were not even considered.

The situation evolved when physicists began to build similar machines to accelerate electrons, most notably the betatron between 1935 and 1940 in Germany and the United States and then, the synchrotron, invented by Vladimir Veksler in 1944 and assembled for the first time by Edwin McMillan in the 1950s. For both types of accelerator, energy losses by radiation were soon understood to be a problem which inhibited the quest for higher energies. In Europe, the great pioneer of electron accelerators was Wolfgang Paul. In the 1950s, he discretely put together a small synchrotron in the basement of the Physikalisches Institut of the Friedrich Wilhelms University in Bonn. He did so discretely because, at least in principle, it was forbidden for German scientists to work on nuclear physics after the Second World War. Fortunately for him, he worked in the British sector, and the officer responsible was a tolerant man, who understood that it was not realistic to hold back the work of one of the greatest physicists of his generation. So, they agreed between themselves not to call the apparatus an accelerator, but to describe it as an 'apparatus for testing electron quadrupole lenses' (which, by the way, had also been invented by Wolfgang Paul). The British officer was quite right. It is stupid to hold back creative researchers.

At the time, Schott's work had been forgotten by almost everyone, including the physicists building these machines. With the invention of the betatron and the discovery of radiative losses, it was another theorist, Julian Schwinger, an American, who stepped in and took it upon himself to explain to the world the origin of what became known subsequently as synchrotron radiation. He devoted an often quoted long and general article to the subject, published in 1949. This form of light – he explained to a new generation of readers who were surprised to discover that accelerated charges emit electromagnetic radiation — originates simply when charged particles are constrained to move on circular orbits. Indeed, as they are free, it follows from Maxwell's theory of electromagnetism that they cannot do otherwise but radiate. This is the mechanism by which they are prevented from reaching the speed of light which, as everybody knows, would be contrary to fundamental principles explained by Albert Einstein.

Everybody was happy with this nice result and it was decided to attribute the discovery of the new type of radiation to Schwinger. Again, poor George Schott had missed the boat, but he was no longer around to complain about it.

If Schwinger's paper had been read a little more carefully, one might have noticed that Schott's name actually did appear in tiny letters in a footnote to the first page. However, as it was not in the main text, nor even in the introduction to the paper, nobody noticed and nobody knew who he was anyway, so why bother to chase up a footnote? It should be added that Schott's treatise on circulating charges, despite the exclusive Adams Prize awarded by the University of Cambridge, had been completely forgotten in the meantime. Indeed, very few copies, if any, were still lying around. Schott's old book had just dropped by the wayside.

George Adolphus Schott spent all his life in relative obscurity and this was not about to change. He had always made a study of elegance and distinction and the first rule to be respected in order to achieve these social qualities is discretion. Not being noticed had been his way of life and he had always hated trumpeting results to the world, especially if they were his own.

As for *his* radiation, his very special form of light, Schott would have been perfectly happy to hear that it still does not carry his name. He would not have found this at all surprising.

Anyway, Schott's radiation soon became an embarrassment for physicists intent on achieving high energies. It was, if the truth be told, a real nuisance. In order to reduce its effect, it became necessary to increase the radius of accelerators as much as possible so as to limit the curvature of the orbit. This led to the construction of huge machines. The only alternative is the linear accelerator, but winding up the energy turn by turn, which is so convenient in a synchrotron, is not possible in this case.

So, machine builders would have cursed Schott anyway at this point in the subject. Everything was done to try and get rid of this awkward radiation and achieve higher energies. Electrons were replaced by protons which, being heavier, emitted less radiation. This was the era of huge accelerators and the birth of high energy physics. Radiative losses, in this context, were retrograde and synchrotron radiation was hidden from view. Attempting to observe it was not in the least bit encouraged.

It was an Armenian physicist, Diran Tomboulian, together with his American colleague Paul Leon Hartman, who finally obtained permission from their high energy physicist colleagues to observe synchrotron radiation for the first time at the accelerator in Cornell in 1953. But, as soon as they had obtained the first sighting of it, their experiment was dismantled to make way for more important things. There was no real interest in this nuisance radiation. So, not only Schott, but also the light he devoted his life to understanding, were both swept under the carpet.

Then came the dawn of a new era, for sociological as well as scientific reasons. There were now accelerators in many places. They had mushroomed during the race for higher and higher energies and were littered over all the developed world. What was to become of them? Some use had to be found for the ones which had been overtaken, which were costly and beautiful toys. One had to find a purpose, not only for the machines themselves, but also for all the specialized teams whose expertise was at the forefront of accelerator physics. To this end, interest in synchrotron radiation was revived and its properties seemed suddenly very important. Better still, it was discovered that there were unique advantages in exploiting it, not only in physics, but also in biology, medicine and in many other fields. Synchrotron radiation was novel. It justified building many different types of accelerators, whose purpose was no longer to achieve high energies for particle physics but was now to maximize and concentrate the emitted radiation for new branches of research.

These new machines were described as synchrotron radiation sources. It was fortunate that they had appeared on the scene, because high-energy particle accelerators were becoming too expensive for national budgets and had become international projects. So, the existence of the synchrotron radiation sources, which accelerated electrons only to lower energy enabled individual countries (provided they were prosperous enough) to preserve their national pride by the construction of sophisticated and advanced facilities. They became a symbol of wealth and power for advanced economies. The Japanese, it turned out, were particularly good at building them and collected more machines than anybody else.

Around these accelerators, it proved possible to build a fan of what were termed 'tangents' or 'beam lines' to guide the light to many different experiments. The property which allowed this to be done was one of the discoveries made entirely theoretically by George Adolphus Schott. He had realized a long time before that, if electrons were accelerated to a speed approaching that of light, the 'Schott' radiation would be emitted in a very narrow pencil sweeping around on a tangent to the orbit ahead of the accelerated particle. This was an immediate consequence of Einstein's special theory of relativity.

Unfortunately, he had discovered this property much too long ago. It was described in his book about the radiation emitted by circulating electrons, but nobody remembered that he, in fact, had been the first to explain this effect. One must always be careful about timing new discoveries.

The first purpose-built synchrotron radiation source to be constructed in France was called LURE ('Laboratoire pour l'Utilisation du Rayonnement Electromagnétique') on the Orsay campus near Paris. At the time, the French Government prided itself on having an independent foreign policy somewhat orthogonal to other European nations, especially as concerned the NATO alliance and the role of the United States in the defence of the West. To emphasize their independence during the Cold War, French authorities were accustomed to invite Soviet visitors to their national laboratories. This was not yet the case in the UK, although in the days of Rutherford, the Russian physicist Pyotr Kapitza, who created the first Low-Temperature laboratory in the USSR, had worked at the Cavendish Laboratory in Cambridge. That was how the Kapitza–Dirac effect came about. What probably put a stop to such exchanges was that, when he returned to Russia, Stalin refused to let him travel again. In effect, he trapped him in his own country, forcing him to work exclusively for the Soviet state.

Amongst the books and papers Pyotr Kapitza brought from Cambridge to Russia was the treatise of George Adolphus Schott on the radiation emitted by circulating electrons, which generations of Soviet physicists read with great interest. When the son of Pyotr Kapitza, Serguei, who was also a physicist, came to Paris to visit the LURE laboratory, he was accompanied by a historian of science, Vassili Maximovich Staretzski, who was very surprised by the description which was given of the birth of synchrotron radiation studies in Western laboratories.

'Why do you always refer to the American Julian Schwinger?' he asked. 'Don't you know about the work of Schott?'

The French scientists were puzzled by this question. Who was Schott? Like all Russian comments, it probably had something to do with official propaganda conditioned by the Cold War and rivalry with the United States. Maybe Schott was one of those mythical great scientists invented by the Soviets in order to counter Western claims of supremacy. Probably, he had never existed, since nobody had heard of him before. Everybody knew that the secret services in Moscow worked overtime on behalf of the Soviet Union inventing false rumours and phoney legends to convince the rest of the world that Bolshevik science, in fact, was vastly superior to capitalist discoveries.

So, Professor Staretzski's question received no reply. Clearly, Schott, despite his German-sounding name, must be one of those Russians born of immigrant ancestors from the opposite side of the Niemen River, an adoptive Russian whose biography had been improved to include improbable discoveries.

Nobody wanted to risk a diplomatic incident by entering into some kind of argument with the Russian delegation or requesting clarification of the Russian professor's remark, since the visit was an official one and the professor was a member of the Academy of Sciences of the Union of Socialist Soviet Republics. Unbeknown to the French scientists, the book Kapitza had brought back with him from Cambridge had been read by Professor Staretzski. Was he one of the few who still knew about the work of George Adolphus Schott? At any rate, he was possibly the last person to mention his name.

An ancient Egyptian proverb asserts that a man is not quite dead as long as people still speak his name. In his afterlife, maybe Schott heard the words of Professor Staretzski. If so, he was surely pleased that somebody still knew about him, even if the professor was neither English nor German.

Schott's name has probably never been mentioned since, except by a few chess players who still recall some of the brilliant moves he invented.

The End

The End

At the end of the day, what can one add? Who of the three great minds who thought hardest about the apparently simple problem of the Planetary Atom — Rutherford, Bohr and Schott — was actually right and who was wrong? Is the problem completely solved today, or does it remain as a long-standing dilemma?

The Atomic Hypothesis

The atomic hypothesis in itself is possibly the most significant contribution physics has made to the understanding of the world around us. The great theorist Feynman, in his celebrated lectures, went further: he described it as the most important idea in the history of science, the single result which should be saved from armageddon in the event of total annihilation, to be passed on to some future civilization.

The hypothesis in itself is extremely ancient. It dates back at least to the pre-Socratic philosopher Democritus of Thrace, to whom the concept that all matter is made of atoms, which are all the same and are its smallest constituents, is usually attributed.

For some reason, commentators of Greek philosophy stop there. However, Democritus actually said somewhat more. Like many great thinkers of

215

Antiquity, he was conscious that he might be the first to express a novel idea in words. So, he was very concerned with complete and proper definitions. Perhaps for this reason, he asked himself the question: what else is there in between the atoms, which distinguishes this space from the rest of reality? And he came up with the brilliant answer: it simply contains no atoms, i.e. no matter at all! He had just described absolute emptiness, or what we might call 'nothingness.' He believed that this strange absence of anything tangible might still, somehow, belong to the real world.

Today, the complete absence of atoms, i.e. the empty space which Democritus placed for the first time in between the atoms is what we call the *vacuum* — a very important concept in physics. From the outset, thanks to Democritus, the history of the vacuum has gone naturally hand in hand with that of the atom.

Does Nature 'Abhor a Vacuum'?

The first person to object to this construct was Aristotle. For some reason, he disliked Democritus and his followers anyway, so he came out with a sweeping statement which seemed to put paid to their whole line of thought. He simply declared: *Nature abhors a vacuum.* Aristotle carried great authority, so the matter was laid to rest by many thinking people (with a few notable exceptions like Epicurus) for centuries. It was replaced by *plenism*: the notion that, somehow, every part of space is packed with something. If this were true, there could be no such thing as atoms with 'empty' spaces between them, and Democritus was necessarily wrong. Plenism was generally accepted, even by scientists. It was consistent with Plato's philosophy because nothingness could hardly possess an essence, so could not be allowed to exist as such. This seemed a self-evident conclusion. It was translated into Latin as *horror vacui*. Aristotle's principle was restated with force by Rabelais in 1530 as *natura abhorret vacuum*, and thus matters stood, with little contradiction, even after the Renaissance.

Further progress depended on observation, and observing atoms — or their absence — is of course very difficult. Observing the vacuum might seem the greater challenge of the two but, strangely, this result came first, in a wonderful experiment conducted involuntarily by Florentine engineers making water pumps. Try as they might to improve the quality of their equipment, they found they could not raise water above a fixed height. The physicist Torricelli, called in to investigate the reason, found no real answer but had the bright idea of replacing water with mercury. He thus reduced the height of the column of liquid to 76 centimetres and, without quite realizing what he had done, invented the barometer.

It fell to Blaise Pascal, the French physicist, mathematician and philosopher, who had developed hydrostatics, to give a fascinating explanation. Torricelli's barometer was an apparatus to weigh the Earth's atmosphere. Above the atmosphere, and also above the column of mercury, there was simply nothing. Improving the pumps would make no difference. The space above the column of mercury was just empty. Here was the vacuum for all to see. So, in fact, Aristotle had been wrong all along. Emptiness was not impossible. Nature did not object to it. Indeed, there was empty space almost everywhere in the universe and, implicitly, this meant that the field was open again for Democritus since the theory of plenism had just been overthrown.

Coincidentally, around the same time, Dutch opticians developed the microscope and observed tiny mites that live in cheese. They were so confident in the quality of their new device that they immediately asserted: these are the smallest animals that can exist.

'Not so!' exclaimed Pascal. Without further prompting, he sat down to write one of the famous pages in the history of philosophy, which complements, but at the same time challenges, the whole basis of atomism in science.

According to Pascal, these animals could certainly not be the smallest because to exist as beings, they must have a nervous system and blood vessels of their own, and these are necessarily all made of atoms. Then, in one giant leap of the mind, he added that the atoms themselves must each be similar to microscopic solar systems, containing planets, within which would be other beings which must involve yet much smaller mites, then more, even smaller atoms and so on and so forth, *ad infinitum*. Pascal had just invented the infinitely divisible universe, in which atoms themselves are not the smallest constituents of matter, but could, in turn, be divided into smaller and smaller pieces, so small that the Dutch microscopes would never be able to perceive them. This idea, in some way, contradicted the concept of Democritus, who saw atoms as the smallest constituents of matter, but seemed to sit more comfortably with the smoothness of space in classical mechanics, as it was developing at the time.

As was pointed out much later by Dirac and others, infinite divisibility also contradicts the foundations of thermodynamics and quantum mechanics, but that is another story.

The next steps were taken by chemists, who recognized that there are different types of matter, implying different kinds of atoms which combine in fixed proportions.

The Quest for 'the *Æther*' and the Birth of Relativity

There, the atom-plus-vacuum theory stood, essentially without change, until Maxwell, while setting up electromagnetic theory, asked a new question: how does light propagate across empty space and reach us from the distant stars? Again, the problem was the very emptiness of vacuum. It was very hard to imagine any energy being transported through 'nothingness' over millions and millions of kilometres. Indeed, it seemed inconceivable. To overcome this difficulty, Maxwell proposed to fill previously empty space with a mysterious substance he baptized the *æther*. This was a bit like a return to 'plenism' except for the fact that the æther was different from matter as such. It was a strange, matter-free 'æthereal' fluid, the only observable function of which was to enable the propagation of light.

Looking for experimental evidence to detect the presence of the æther was the purpose of a famous experiment by Michelson and Morley, using the optical interferometer Michelson had just invented, but the results were very disappointing. After many frustrating attempts, the experimenters were forced to conclude that there was no observable evidence for its existence. So, while the mathematics of Maxwell's theory could not be doubted, the experiments suggested that the æther is not necessary for light to propagate. Thus, the vacuum again became 'empty' as before.

The Michelson–Morley experiment left physicists in a deep quandary about the laws of Newtonian mechanics, which could not be reconciled with the fact that the maximum velocity attainable in any direction is given by the speed of light, regarded at that time as strictly constant at any point in space. It was only when Lorentz introduced his transformation and when Einstein formulated the special theory of relativity that the inconsistency between mechanics and electromagnetic theory, revealed by experiment, was resolved, at least from the mathematical point of view.

Meanwhile, attention was turning to atoms, which could be detected and studied, both by using light and by new tools the physicists were mastering: Rutherford had also begun to develop nuclear physics.

Early steps in this new subject led to Rutherford's backscattering experiment, which established that mass is concentrated in the centre of atoms. This observation led Rutherford to propose his 'planetary' model of the atom, subsequently refined by Bohr, who introduced the so-called quantum or 'stationary' states.

An implication of Rutherford's model was that the vacuum could no longer be completely external to the atom. Vacuum could now exist also within the atom, between the nucleus and the 'point-like' electrons, both of which contain matter. So, the definition of the vacuum, according to Democritus, needed to be refined. It was no longer merely the absence of atoms but also the absence of any kind of particle-containing matter.

Schott Enters the Scene

By this time, it had become accepted that one should do away completely with Maxwell's æther, since it was not observable, and that electromagnetic radiation could nonetheless propagate in 'free space' although the mechanism by which this occurs remained very mysterious. Schott was aware of this conclusion. He took on board that electromagnetic radiation is capable of propagating in a vacuum. By rights, therefore, Maxwell's theory should apply everywhere, including inside atoms, and, therefore, the energy of the electrons, on circular orbits according to Rutherford's model, should be radiated away, leading them inevitably to 'collapse' onto the nucleus. So, according to Schott, Rutherford's atom could not exist. It would be unstable.

In a subtle way, all three protagonists in *The Planetary Atom* (Rutherford, Bohr and Schott) were somehow correct in their conclusion and, in equally subtle ways, all three of them were also wrong. Rutherford was right that the mass of the atom is mainly concentrated in its centre and that nearly all of it lies in the nucleus. But he was wrong to think that a simple planetary model could account for all atomic properties without addressing the issue of the electromagnetic field. Bohr was right in understanding that atoms are stable in their ground state, while excited states are governed by a new kind of mechanics involving quantization. However, he was wrong, as turned out later, in imagining that this new mechanics could be set up merely by arbitrarily turning off electromagnetic radiation. He was lucky that his approximation is almost perfect when it comes to calculating atomic properties, but his representation of the planetary atom (the Bohr model) is neither fully consistent nor fully correct, for reasons which Schott had grasped. Even in the next step beyond the Bohr model (the Schrödinger equation on which elementary quantum mechanics is based), the radiation field is not included and electromagnetism is not treated properly.

Dirac took matters one step further in setting up his own equation by attempting to satisfy the new rules of atomic physics together with special relativity or, at least, with the Lorentz transformations, which incorporate a crucial feature of Maxwell's electromagnetism: the constancy of the speed of light. However, by his own reckoning, he still did not achieve a fully relativistic theory. As he put it, the Dirac equation could not be properly relativistic, as it remained a single-particle equation.

Why Worry About Relativity?

In the background of the whole argument lurks the question: how do the new mechanics Bohr had somehow introduced into the problem and all the refinements added to it by others sit with the theory of relativity? The fact is that they remain uncomfortable partners, despite the wonderful advance made by Dirac from which the property known as spin suddenly emerges, as though by mathematical magic. This incompatibility of two great theories is why Einstein, all along, was implicitly involved in the whole debate. He had his own reasons for remaining aloof from quantum theory. His main contribution to the new mechanics was to suggest that, if it applied so successfully to particles, it should also, for consistency, be applied to radiation. This requires the introduction of the discrete bursts of radiation he termed 'photons'. Somewhat paradoxically, it was not for relativity (either special or general) that he was awarded the Nobel prize, but for his introduction of photons into quantum physics.

Hidden within the new quantum mechanics was a new difficulty. Would quantization imply that the nature of empty space itself is discrete, with a difference between space and time written into the basic equations, or is space-time truly four-dimensional, smooth, infinitely divisible, in the manner imagined by Pascal, and continuous as Einstein insisted must be the case, away from the singularities we call black holes today?

Schott Neglected the Photon

On one issue, Schott was quite right: the inclusion of electromagnetic radiation, which plays such an important part in relativity, is essential also in the description of atoms, but he was wrong in not considering the existence of Einstein's photons (what is called today 'second quantization'). He was right again in deducing the properties of the, as yet undiscovered, synchrotron radiation emitted by particle accelerators. He had formulated a brilliant application of Maxwell's theory of electromagnetism, but, again, did not envisage the quantization of light. Like Bohr (but at the opposite extreme in theoretical terms), Schott was also lucky: he had found an approximation which is almost exactly correct for the radiation emitted by accelerators. The existence of photons has very little influence at all on the problem, for most situations.

It cannot be completely neglected, however. It turns out to have two main consequences. The first is related to momentum conservation. Every time a photon is ejected by an electron in a circular orbit, there is a small recoil. As there are many photons emitted randomly by one accelerated electron in a single turn, the result is to broaden out the electron beam, which can never be on an ideally circular path, even in an ideal machine, when photons are emitted. This 'fuzzes out' the orbit, which, in turn, is why synchrotron radiation can never achieve infinite brightness. The second arises in a machine known as a storage ring, where the electron beam, after being accelerated to a given energy, is maintained or 'stored' and allowed to fade away slowly, electron by electron, as it radiates away its energy. Again, in an ideally adjusted machine with a perfect vacuum, electrons would have to disappear from their orbit in steps, owing to the emission of energy packets, each one corresponding to a photon. These exceedingly small effects remain mostly unobservable. Thus Schott's theory, which does not allow for them, is perfectly adequate to account for nearly all situations.

It remains today that all three protagonists, Rutherford, Bohr and Schott, made brilliant contributions to physics, laying the foundations for subsequent advances. Schott, for various reasons, has been forgotten. It is only fair to set the record straight today by including him, because the experimental observation of synchrotron radiation (it should really be called 'Schott radiation') brought confirmation that his criticism of Rutherford's model was well-founded and cannot be disregarded.

As often happens in science, this debate about fundamentals (which is not completely over) should not be treated as an open and shut case. In fact, it underpins the ongoing subject of quantum optics — a rich field of study today. One of the themes of this area of physics has been the quest for incontrovertible experimental evidence in favour of photons. This is surprisingly difficult to achieve, but has now been satisfactorily resolved in the so-called 'random telegraph' and 'photon antibunching' experiments. This being said, the complete description of a real atom in a quantized radiation field remains a difficult problem even today.

What Does All This Do to the Vacuum?

Likewise, the vacuum, which was created empty by Democritus, then filled with the æther by Maxwell and emptied again after Michelson and Morley, has become a singularly complex object of study in modern times. For a start, Dirac, as a consequence of his equation, filled it with so-called 'negative-energy states' of the electron. This extraordinary idea had the benefit of leading to the discovery of the positron (the antiparticle of the electron) — an experimental discovery in which Patrick Blackett played a key role. However, antiparticles came in large numbers every time a particle was discovered. The great theorist Majorana was concerned about this problem. He sought particles which could be their own antiparticles. However, they were not the rule. The number of antiparticles grew, causing such a bottleneck that the vacuum became a new difficulty, described by some as the 'dustbin' of theoretical physics.

The question of the propagation of photons through a completely 'empty' vacuum also remained something of a mystery. Quantum mechanics allowed the issue of the æther to be revived mathematically by seeking a model for the propagation of photons involving combinations of quantized oscillators, hidden from view most of the time by the uncertainty principle.

As if this were not enough, general relativity opened up new complexities in the form of singularities in space-time or 'black holes', surrounded by many strange properties of 'empty' space, new forms of radiation involving gravitation, and raising the frightening possibility of tunnelling through into parallel universes to our own, as described with much relish by Hawking…

Conclusion

Democritus would surely have been astonished by the extraordinary transformation his simple, straightforward and empty vacuum has undergone. Indeed, if one takes on board the results of quantum theory, it becomes debatable whether a true 'vacuum' can really exist in physics. In a strange way, the problem has bounced back and forth over the centuries. Perhaps, after all, Aristotle had a point. Nature may not allow a complete and utter vacuum to exist, and the first manifestation of this is indeed through the propagation of light.

So, there is no such thing as 'the end of the day' imagined in my opening sentence. The days have no end for those who really seek the truth. Even the humble problem of the planetary atom turns into one of the great debates of modern science. It remains one of the keys to our understanding of matter and provided a fundamental step in intellectual progress during the twentieth century.

Jean-Patrick Connerade
Imperial College London
March 2021

Further Reading

In English:

Bohr, Niels
The Theory of Spectra and Atomic Constitution,
Cambridge University Press, Cambridge, UK, 1922

Connerade, Jean-Patrick
Highly Excited Atoms,
Cambridge University Press, Cambridge, UK, 1998

Dirac, Paul
The Principles of Quantum Mechanics,
4th Edition, Oxford University Press, Oxford, UK, 1958

Einstein, Albert
Relativity: The Special and General Theory,
Dover Books on Physics, Dover Publications,
New York, USA, 2010;
Ideas and Opinions,
Crown Publishers Inc, New York, USA, 1982

Feynman, Richard
The Feynman Lectures on Physics
with Robert P. Leighton & Matthew Sands

Published by Addison-Wesley Publishing Company
for California Institute of Technology 1975–1976,
Reading, Massachusetts, USA, 1975

Hawking, Stephen W
A Brief History of Time: From the Big Bang to Black Holes, Bantam
Books, New York, USA, 1988

Ryder, Lewis
Quantum Field Theory, Cambridge University Press,
Cambridge, UK, 1985

Saleem, Mohammad and Rafique, Muhammad
Special Relativity, Ellis Norwood, New York, 1992

Schott, George Adolphus
Electromagnetic Radiation and the Mechanical Reactions Arising from It,
Cambridge University Press, Cambridge, UK, 1912

Schutz, Bernard
A First Course in General Relativity,
Cambridge University Press, Cambridge, UK, 1985

Series, George
The Spectrum of Atomic Hydrogen,
World Scientific Publishing Co., Singapore, 1988

Wu, Ta-You
Quantum Mechanics,
World Scientific Publishing Co., Singapore, 1986

In French:

Cauchois, Yvette et Héno, Yvonne
Cheminement des particules chargées,
Gauthier-Villars, Paris, France, 1964

Curie, Eve
Madame Curie, Librería Hachette, SA, Buenos Aires, Argentina, 1944

de Broglie, Louis
L'électron magnétique (*théorie de Dirac*),
Hermann et Cie, Editeurs, Paris, 1934;
Une tentative d'interprétation causale et non linéaire de la mécanique ondulatoire: la théorie de la double solution, Gauthier-Villars, Paris, France, 1956;
Étude critique des bases de l'interprétation actuelle de la mécanique ondulatoire, Gauthier-Villars, Paris, France, 1963

Perrin, Jean
Grains de matière et de lumière,
Hermann & Cie, Editeurs, Paris, France, 1935

Printed in the United States
by Baker & Taylor Publisher Services

Printed in the United States
by Baker & Taylor Publisher Services